CONCEPTS OF MEMBRANES IN REGULATION AND EXCITATION

Concepts of Membranes in Regulation and Excitation

Edited by

M. Rocha e Silva
Department of Pharmacology
Faculty of Medicine of Ribeirão Preto
University of São Paulo
Ribeirão Preto, Brazil

Guilherme Suarez-Kurtz
Department of Pharmacology
Institute of Biomedical Sciences
Federal University of Rio de Janeiro
Rio de Janeiro, Brazil

Raven Press, Publishers ▪ New York

WITHDRAWN FROM
YALE UNIV LIBRARY
AUG 2 0 2010

Foreword

This volume assembles the work of scientists of widely differing disciplines but with a common interest in membrane-associated phenomena. The approaches and techniques applied in their studies of membrane-related processes reflected the diverse backgrounds of the contributing scientists. The chemists and biophysicists defined membrane permeabilities in terms of electron spin resonance data, phase-plane analysis, and voltage-clamp measurements. The physiologists and biochemists investigating membrane excitatory and regulatory processes monitored specific metabolic steps, ionic fluxes, and tension. The pharmacologists employed drugs and toxins as probes for studying drug-receptor reactions as well as membrane conformational changes. Immunological techniques were also brought to bear and in combination with electrophysiological methodology were used to follow membrane–membrane interactions.

Calcium was found to play an integral role in many of the processes that were studied, including (a) membrane excitatory signaling events, (b) modulation of cell–cell and drug-receptor interactions, and (c) membrane conformational changes.

The discussion of concepts of membrane phenomena by the multidisciplined group of authors was highly stimulating. Both the organizing committee and the Brazilian Academy of Sciences are to be commended for arranging and sponsoring this unique and valuable symposium which was held in Rio de Janeiro in June, 1974.

John P. Reuben
Departments of Physiology and Neurology
College of Physicians and Surgeons
Columbia University
New York, New York

Acknowledgments

We are greatly indebted to Dr. Hanna Rothschild for her invaluable assistance in the editing of this volume. We wish to express our gratitude to our colleagues of the organizing committee of the symposium on which this volume is based, Drs. A. Antonio, A. P. Corrado, C. R. Diniz, and A. P. Leao. It is a pleasure to acknowledge the collaboration of Dr. John R. Reuben throughout the organization of the meetings. The symposium was supported by special grants from the Brazilian Academy of Sciences, the National Research Council of Brazil, and Fundação do Amparo a Pesquisa do Estado de São Paulo.

M. Rocha e Silva
G. Suarez-Kurtz
(*May 1975*)

Contents

1 Calcium Fluxes in Skeletal Muscle and Integration of Metabolic and Contractile Events
C. Paul Bianchi

7 ATP $\rightleftharpoons$ P$_i$ Exchange Catalyzed by Sarcoplasmic Reticulum with and without a Ca^{++} Concentration Gradient
L. de Meis, Maria da Glória Costa Carvalho, and Martha M. Sorenson

21 Ca Regulation in "Skinned" Muscle Fibers
John P. Reuben, Donald S. Wood, Jenny R. Zollman, and Philip W. Brandt

41 Is a Ca Current Necessary for Excitation-Contraction Coupling in Skeletal Muscle?
Guilherme Suarez-Kurtz and John P. Reuben

55 Some Characteristics of the Excitation-Contraction Coupling Process in Smooth Muscle
Leon Hurwitz

73 Calcium Translocation in Rat Vas Deferens in the Light of Receptor Theory
A. Jurkiewicz, R. P. Markus, and Z. P. Picarelli

85 Phase-Plane Determination of Membrane Currents in Propagated Action Potentials: Possibilities and Difficulties
Antonio Paes de Carvalho

97 Active Sodium Transport and Oxygen Coupling in Amphibian Skin
F. Lacaz Vieira and G. Danisi

107 Hydrogen Ion Transport in Renal Tubules: Effect of Acetazolamide
G. Malnic, Margarida de Mello Aires, and A. C. Cassola

125 Studies on the Process of Acidification in the Urinary Bladder of the Toad *In Vitro*
M. R. Furtado and A. C. Nero

135 Intercellular Communication in Lymphocytes
Gilberto M. de Oliveira-Castro, Marcello A. Barcinski, and Ionice F. Gaziri

CONTENTS

145   A Model for Angiotensin Receptors in Smooth Muscle Cells
A. C. M. Paiva and Therezinha B. Paiva

155   Permeability Characteristics of Lipid Bilayers Revealed by Spin Probes
Shirley Schreier-Muccillo, D. Marsh, and I. C. P. Smith

167   Angiotensin Receptors in Smooth Muscle Cell Membranes: A Spin-label Study
Mecia M. Oliveira, Shirley Schreier-Muccillo, Suma Shimuta, Gregory Niculitcheff, and A. C. M. Paiva

181   A Thermodynamic Approach to Problems of Drug Antagonism: The Microphysical Model
M. Rocha e Silva

193   Neurotransmitter Release by the Toxin of Brazilian Scorpion Venom (*Tityus serrulatus* Lutz e Mello)
A. P. Corrado, C. R. Diniz, and A. Antonio

201   Competitive Antagonism between Calcium and Aminoglycoside Antibiotics in Skeletal and Smooth Muscles
A. P. Corrado, W. A. Prado, and I. Pimenta de Morais

217   Biochemical Properties of *Tityrus* Toxin
C. R. Diniz, J. Coutinho Netto, A. F. Pimenta, and R. E. Larson

223   *Index*

Contributors

A. ANTONIO, *Department of Pharmacology, Faculty of Medicine of Ribeirão Preto, University of São Paulo, Ribeirão Preto, Brazil 14100*

MARCELLO A. BARCINSKI, *Institute of Biophysics, Federal University of Rio de Janeiro, Rio de Janeiro, Brazil 20000*

C. PAUL BIANCHI, *Department of Pharmacology, School of Medicine, University of Pennsylvania, Philadelphia, Pennsylvania 19174*

PHILIP W. BRANDT, *Department of Anatomy, College of Physicians and Surgeons, Columbia University, New York, New York 10032*

MARIA DA GLÓRIA COSTA CARVALHO, *Institute of Biophysics, Federal University of Rio de Janeiro, Rio de Janeiro, Brazil 20000*

ANTONIO C. CASSOLA, *Department of Physiology, Institute of Biomedical Sciences, University of São Paulo, São Paulo, Brazil 01000*

A. P. CORRADO, *Department of Pharmacology, Faculty of Medicine of Ribeirão Preto, University of São Paulo, Ribeirão Preto, Brazil 14100*

J. COUTINHO NETTO, *Laboratory of Neurochemistry, Department of Biochemistry, Faculty of Medicine of Ribeirão Preto, University of São Paulo, Ribeirão Preto, Brazil 14100*

G. DANISI, *Department of Physiology and Pharmacology, Institute of Biomedical Sciences, University of São Paulo, São Paulo, Brazil 01000*

C. R. DINIZ, *Department of Biochemistry, Faculty of Medicine of Ribeirão Preto, University of São Paulo, Ribeirão Preto, Brazil 14100*

M. R. FURTADO, *Department of Internal Medicine, Faculty of Medicine of Ribeirão Preto, University of São Paulo, Ribeirão Preto, Brazil 14100*

Ionice F. GAZIRI, *Institute of Biophysics, Federal University of Rio de Janeiro, Rio de Janeiro, Brazil 20000*

LEON HURWITZ, *Department of Pharmacology, University of New Mexico School of Medicine, Albuquerque, New Mexico 87131*

A. JURKIEWICZ, *Department of Biochemistry and Pharmacology, Escola Paulista de Medicina, São Paulo, Brazil 04023*

F. LACAZ VIEIRA, *Department of Physiology and Pharmacology, Institute of Biomedical Sciences, University of São Paulo, São Paulo, Brazil 01000*

R. E. LARSON, *Laboratory of Neurochemistry, Department of Biochemistry, Faculty of Medicine of Ribeirão Preto, University of São Paulo, Ribeirão Preto, Brazil 14100*

G. MALNIC, *Department of Physiology, Institute of Biomedical Sciences, University of São Paulo, São Paulo, Brazil 01000*

L. DE MEIS, *Institute of Biophysics, Federal University of Rio de Janeiro, Rio de Janeiro, Brazil 20000*

R. P. MARKUS, *Department of Biochemistry and Pharmacology, Escola Paulista de Medicina, São Paulo, Brazil 04023*

D. MARSH, *Department of Biochemistry, University of Oxford, Oxford, England*

A. C. NERO, *Department of Internal Medicine, Faculty of Medicine of Ribeirão Preto, University of São Paulo, Ribeirão Preto, Brazil 14100*

GREGORY NICULITCHEFF, *Department of Biochemistry, Faculty of Medicine of Ribeirão Preto, University of São Paulo, Ribeirão Preto, Brazil 14100*

MECIA, M. OLIVIERA, *Department of Biophysics and Physiology, Escola Paulista de Medicina, São Paulo, Brazil 04023*

GILBERTO DE OLIVEIRA-CASTRO, *Institute of Biophysics, Federal University of Rio de Janeiro, Rio de Janeiro, Brazil 20000*

ANTONIO PAES DE CARVALHO, *Institute of Biophysics, Federal University of Rio de Janeiro, Rio de Janeiro, Brazil 20000*

C. M. PAIVA, *Department of Biophysics and Physiology, Escola Paulista de Medicina, São Paulo, Brazil 04023*

THEREZINHA PAIVA, *Department of Biophysics and Physiology, Escola Paulista de Medicina, São Paulo, Brazil 04023*

I. PIMENTA DE MORAIS, *Department of Pharmacology, Faculty of Medicine of Ribeirão Preto, University of São Paulo, Ribeirão Preto, Brazil 14100*

W. A. PRADO, *Department of Pharmacology, Faculty of Medicine of Ribeirão Preto, University of São Paulo, Ribeirão Preto, Brazil 14100*

JOHN P. REUBEN, *Departments of Physiology and Neurology, College of Physicians and Surgeons, Columbia University, New York, New York 10032*

M. ROCHA E SILVA, *Department of Pharmacology, Faculty of Medicine of Ribeirão Preto, University of São Paulo, Ribeirão Preto, Brazil 14100*

SHIRLEY SCHREIER-MUCCILLO, *Department of Biochemistry, Institute of Chemistry, University of São Paulo, São Paulo, Brazil 01000*

SUMA SHIMUTA, *Department of Biochemistry, Institute of Chemistry, University of São Paulo, São Paulo, Brazil 01000*

I. C. P. SMITH, *Division of Biological Sciences, National Research Council, Ottawa, Canada*

MARTHA M. SORENSON, *Institute of Biophysics, Federal University of Rio de Janeiro, Rio de Janeiro, Brazil 20000*

GUILHERME SUAREZ-KURTZ, *Department of Pharmacology, Institute of Biomedical Sciences, Federal University of Rio de Janeiro, Rio de Janeiro, Brazil 20000*

DONALD S. WOOD, *Department of Anatomy, College of Physicians and Surgeons, Columbia University, New York, New York 10032*

JENNY R. ZOLLMAN, *Department of Anatomy, College of Physicians and Surgeons, Columbia University, New York, New York 10032*

Concepts of Membranes in Regulation and Excitation,
edited by M. Rocha e Silva and G. Suarez-Kurtz.
Raven Press, New York © 1975.

Calcium Fluxes in Skeletal Muscle and Integration of Metabolic and Contractile Events

C. Paul Bianchi

Department of Pharmacology, School of Medicine, University of Pennsylvania, Philadelphia, Pennsylvania 19174

OXYGEN UPTAKE AND CALCIUM FLUXES

The integration of metabolic processes with skeletal muscle function is regulated primarily by the alteration of membrane potential and associated changes in calcium metabolism. In the resting state, the membrane potential of skeletal muscle fiber is high (−90 mV), the muscle is relaxed (sarcoplasmic calcium $< 10^{-7}$ M), and oxygen consumption, or heat production, is low. The oxygen consumption of isolated resting muscle is 1.75 μmole/g/hr, and resting calcium influx at a normal external K^+ concentration of 2.5 mM is 0.102 μmole/g/hr. If the external K^+ concentration of the Ringer's solution bathing the muscle fiber is raised to 20 mM, the membrane potential of the fiber falls to −55 mV. Associated with membrane depolarization is a marked increase in oxygen consumption, which is sustained for at least 60 min. Calcium influx rises from 0.102 μmole/g/hr to 0.132 μmole/g/hr and is also sustained. The maximum increase in oxygen consumption occurs at a depolarization of −48 mV, i.e., at the threshold level for muscle contraction (Hill and Howorth, 1957; Van der Kloot, 1967; and Bianchi, 1968).

At greater levels of membrane depolarization, produced by increasing external K^+ to 80 mM, a phasic contracture occurs, lasting 30 to 60 sec, and is associated with a transient increase in calcium influx, amounting to 11.4 nmoles/g within 30 sec. Oxygen consumption rises to a peak level and after 10 to 15 min falls off to near basal level by 25 min, even though the membrane is maintained in a depolarized state by the continued presence of 80 mM K^+. Calcium influx falls from a peak level of 1.37 μmoles/g/hr (first 30 sec) to a level of 0.078 μmole/g/hr. The striking relationship between calcium influx and oxygen consumption was first noted by Novotny and Vyskocil (1966). They proposed that the extra oxygen consumption caused by potassium depolarization or the application of caffeine was related to calcium release from a bound form.

Increased oxygen uptake due to membrane depolarization is prevented by the removal of calcium from the medium or by the use of agents that prevent the rise in calcium influx, i.e., procaine or barbiturates. High concentrations (i.e., > 5 mM) of divalent ions (Ca^{++}, Mn^{++}, Mg^{++}, Co^{++}) also depress potassium stimulation of oxygen uptake. The inhibitory action of high calcium concentrations depends on the membrane potential. Calcium at a concentration of 12 mM, while inhibiting the rise in oxygen uptake induced in frog sartorius muscle by 20 mM K^+, restores oxygen uptake when added to muscles previously depolarized by 50 mM K^+; i.e., 50 mM K^+ causes an increase in oxygen uptake that peaks within 15 min and then falls back to basal level within 35 min. Van der Kloot (1967) found that raising the external Ca^{++} from 2.5 mM to 12 mM markedly increased oxygen uptake. Van der Kloot (1969) also showed that Ni^{++} could replace Ca^{++} and maintain potassium stimulation of oxygen uptake. In the same study he showed that disruption of the transverse tubule by glycerol treatment (400 mM glycerol + Ringer's solution for 60 min, followed by 60 min recovery in Ringer's solution) prevented the potassium stimulation of oxygen uptake as well as the potassium contracture.

GLYCEROL-TREATED MUSCLES

Table 1 compares calcium influx and oxygen uptake in normal and glycerol-treated muscles. Glycerol treatment does not alter the basal oxygen uptake: 1.75 μmoles O_2/g/hr for normal sartorius muscle (winter frogs) compared with 1.75 μmoles O_2/g/hr for glycerol-treated muscles. The membrane potential of these two preparations is approximately -90 mV. Glycerol treatment increases calcium influx sevenfold, without increasing the basal rate of oxygen consumption. Potassium depolarization leads to in-

TABLE 1. *Calcium influx and oxygen consumption in normal (N) and glycerol-treated (GT) muscles*

	Treatment	K_o^+ (mM)	Calcium influx (μmole/g/hr)	Oxygen uptake (μmole/g/hr)
Winter frogs[a]	N	2.5	0.102	1.75 ± 0.12 (25)
	GT	2.5	0.738	1.75 ± 0.13 (11)
	N	20	0.134	4.92 ± 0.51 (8)
	N	2.5 + 10^{-6} M epinephrine		0.81 ± 0.09 (4)
	N	20 + 10^{-6} M epinephrine		1.72 ± 0.28 (4)
	N	2.5 + 2 mM caffeine	0.217	11.33 ± 0.77 (9)
	N	2.5 + 5 mM caffeine	0.275	10.82 ± 1.00 (7)
Summer frogs	N	2.5		3.33 ± 0.09 (7)
	GT	2.5		3.41 ± 0.34 (12)
	GT	20 mM K		3.71 ± 0.37 (6)
	GT	2.5 + 0.5 mM caffeine		3.26 ± 0.46 (6)
	GT	20 mM K + 0.5 mM caffeine		5.42 ± 0.56 (6)[b]

[a] Winter frogs are more sensitive to K^+ stimulation of oxygen uptake.
[b] Increased oxygen uptake (experimental-control) = 2.48 ± 0.58, $p < 0.01$.

TABLE 2. *Effect of agents that enhance or depress oxygen uptake in frog sartorius muscle on adenyl cyclase activity (% conversion of ^{3}H-adenine to cyclic AMP)*

	Control	Isoproterenol (10^{-5} M)	Epinephrine (10^{-5} M)
% Conversion	0.21 ± 0.02	0.41 ± 0.03	0.41 ± 0.05
% Oxygen uptake	100%	—	60%
	20 mM K_o^+	110 mM K_o^+	1 mM caffeine
% Conversion	0.23 ± 0.04	0.21 ± 0.02	0.13 ± 0.01
% Oxygen uptake	250%	100%	186%

creased calcium influx without increased oxygen uptake. Potassium stimulation of oxygen uptake can be restored by dibutyryl cyclic AMP (Dawson and Bianchi, 1975) or by pretreatment with 0.5 mM caffeine, a concentration of caffeine that by itself has no effect on oxygen uptake. Epinephrine, which increases cyclic AMP levels in muscle, depresses resting oxygen uptake in normal muscle and markedly reduces potassium stimulation of oxygen uptake (Table 1). The restoration of potassium stimulation of oxygen uptake in glycerol-treated muscles by dibutyryl cyclic AMP or low levels of caffeine (<1.0 mM) suggests that in addition to calcium influx or release a cyclic nucleotide is required as a cofactor for the response.

Table 2 compares the effects of agents that enhance or depress oxygen uptake in frog sartorius muscle and their effects on adenyl cyclase activity. Epinephrine depresses oxygen uptake and doubles adenyl cyclase activity. K^+ (20 mM) and caffeine enhance oxygen uptake, but they either have no effect on or depress adenyl cyclase activity. The data in Table 2 suggest that the cyclic nucleotide involved as a second messenger for potassium stimulation of oxygen uptake in sartorius muscle is not cyclic AMP. At present we are measuring changes in the ratio of cyclic GMP to cyclic AMP as a function of K^+ or caffeine to determine if an increase in that ratio is associated with stimulation of oxygen uptake.

CAFFEINE AND MUSCLE METABOLISM

Caffeine causes a marked rise in calcium influx as well as efflux; the calcium efflux is slightly higher than the influx as it does not depend on calcium for calcium exchange (Bianchi, 1961). Although caffeine invariably increases calcium efflux, the rise in efflux can occur without contracture. Thus the most sensitive response of muscle to caffeine is increased oxygen uptake and increased calcium efflux. Contracture occurs at higher levels of caffeine. The range of caffeine concentration required for these effects is narrow: thus 0.5 mM to 0.8 mM caffeine may have no effect on oxygen uptake; 1 to 2 mM caffeine causes a marked increase in calcium exchange, with either no contracture or only a slight contracture. At concentrations greater than 5.0 mM a contracture occurs. The ability of caffeine to induce calcium efflux without a contracture suggests that skeletal muscle possesses a well-developed regulatory system not only for maintaining sarcoplasmic calcium below 10^{-7} M (muscle relaxed) but also for transporting calcium out of the muscle. The increase in oxygen consumption may be related more to calcium efflux or transport from the muscle than to calcium influx. This would be in keeping with (a) the ability of Ni^{++} to replace external calcium and restore potassium stimulation of oxygen uptake, and (b) the ability of caffeine to stimulate oxygen uptake in the absence of external calcium. Agents that block caffeine stimulation of respiration (procaine or procainamide) also block caffeine-induced efflux of calcium.

MALIGNANT HYPERTHERMIA

Lidocaine enhances caffeine-induced contracture and also enhances caffeine stimulation of oxygen uptake and does not block caffeine-induced calcium efflux (Novotny and Bianchi, 1967; Bianchi and Bolton, 1967).

The latter findings lead to the suggestion that volatile anesthetics—especially halothane, which is associated with malignant hyperthermia—might act in a manner similar to lidocaine's (Strobel and Bianchi, 1971; Bianchi, 1973). In a follow-up of these studies, Strobel found that volatile anesthetics at 1 minimum alveolar concentration (MAC) for anesthesia could be classified according to their ability to potentiate a near-threshold caffeine contracture (Table 3). Chloroform and halothane are the most potent, whereas diethyl ether, fluroxene, and nitrous oxide are the least effective in potentiating the near-threshold caffeine contracture.

Malignant hyperthermia in response to general anesthesia is responsible for anesthetic deaths in apparently healthy patients with an incidence of 1 in 15,000 (Britt and Kalow, 1970a,b). Anesthetic hyperpyrexia is not a homogeneous syndrome; in 70% of the cases hyperpyrexia occurs with muscle rigidity, and in 30% of the cases hyperpyrexia occurs without mus-

TABLE 3. *Muscle tensions developed in response to 2 mM caffeine in the presence of 1 MAC anesthetic*

Anesthetic	Concentration (%)	Tension with anesthetic[a] / Tension 2 mM caffeine
Chloroform	0.62 ± 0.06	15.50 ± 3.59
Halothane	0.71 ± 0.13	11.64 ± 2.09
Methoxyflurane	0.19 ± 0.01	10.49 ± 1.46
Cyclopropane	9.2 ± 1.4	5.73 ± 1.78
Enflurane	1.87 ± 0.06	4.29 ± 1.27
Isoflurane	1.18 ± 0.20	3.16 ± 1.13
Diethyl ether	1.64 ± 0.25	1.57 ± 0.30
Fluroxene	3.44 ± 0.34	1.40 ± 0.30
Nitrous oxide	80.42 ± 1.2	1.34 ± 0.23

[a] Ratio given as mean $\pm$ SD ($n = 6$).

cle rigidity. The predisposition to anesthetic hyperpyrexia is linked to a familial abnormality, perhaps a supersensitivity of the sarcoplasmic reticulum to caffeinelike drugs. The rise in body temperature without muscle rigidity is similar to the ability of low levels of caffeine to increase calcium efflux and oxygen uptake (heat production) without causing muscle contracture. In order to elucidate this syndrome, we must determine what energy sink (calcium transport or uncoupling of an ATP-dependent process) is responsible for increasing oxygen uptake in muscle while the muscle is neither doing mechanical work nor recovering from previous mechanical work. Once the mechanism is understood, we may be able to relate the process to a functional role of skeletal muscle in the maintenance of basal body heat.

SUMMARY

In summary, we know that calcium and a cyclic nucleotide are required for stimulation of oxygen uptake as a function of membrane depolarizations between -90 mV and -50 mV. Membrane depolarization of greater magnitude leads to a phasic contracture and a switch from aerobic to anaerobic metabolism. During recovery from potassium-induced contracture, metabolism becomes primarily aerobic, calcium influx and efflux decrease, and it is during this period of sustained membrane depolarization to levels greater than -50 mV that increasing external calcium enhances rather than decreases oxygen uptake. Further experimental work will help us to understand the metabolic reactions involved in these processes and their regulation by calcium and cyclic nucleotides in relation to membrane potential.

REFERENCES

Bianchi, C. P. (1961): The effect of caffeine on radiocalcium movement in frog sartorius. *J. Gen. Physiol.* 44:845.
Bianchi, C. P. (1968): *Cell Calcium.* Butterworths, London.
Bianchi, C. P. (1973): Cell calcium and malignant hyperthermia. In: *International Symposium on Malignant Hyperthermia*, edited by R. A. Gordon, B. A. Britt, and W. Kalow, pp. 147–151. Charles C. Thomas, Illinois.
Bianchi, C. P., and Bolton, T. C. (1967): Action of local anesthetics on coupling systems in muscle. *J. Pharmacol. Exp. Ther.* 157:388.
Britt, B. A., and Kalow, W. (1970*a*): Malignant hyperthermia: A statistical review. *Can. Anaesth. Soc. J.* 17:293.
Britt, B. A., and Kalow, W. (1970*b*): Malignant hyperthermia: Aetiology unknown! *Can. Anaesth. Soc. J.* 17:316.
Dawson, M. J., and Bianchi, C. P. (1975): Restoration of potassium-stimulated respiration of "glycerol-treated" muscle. *Eur. J. Pharmacol. In press.*
Hill, A. V., and Howorth, J. V. (1957): The effect of potassium on the resting metabolism of the frog's sartorius. *Proc. Roy. Soc. (London) B* 147:21.
Novotny, I., and Bianchi, C. P. (1967): The effect of xylocaine on oxygen consumption in the frog sartorius. *J. Pharmacol. Exp. Ther.* 155:456.
Novotny, I., and Vyskocil, F. (1966): Possible role of Ca ions in the resting metabolism of frog sartorius muscle during potassium depolarization. *J. Cell. Physiol.* 67:159.

Strobel, G. E., and Bianchi, C. P. (1971): An in-vitro model of anesthetic hypertonic hyperpyrexia, halothane–Caffeine-induced muscle contractures: Prevention of contracture by procainamide. *Anesthesiology* 35:465.

Van der Kloot, W. C. (1967): Potassium stimulated respiration and intracellular calcium release in frog skeletal muscle. *J. Physiol.* 191:141.

Van der Kloot, W. C. (1969): The steps between depolarization and the increase in the respiration of frog skeletal muscle. *J. Physiol.* 204:551.

Concepts of Membranes in Regulation and Excitation,
edited by M. Rocha e Silva and G. Suarez-Kurtz.
Raven Press, New York © 1975.

ATP $\rightleftharpoons$ P$_i$ Exchange Catalyzed by Sarcoplasmic Reticulum with and without a Ca^{++} Concentration Gradient

L. de Meis, Maria da Glória Costa Carvalho,
and Martha M. Sorenson*

*Institute of Biophysics, Center for Medical Sciences, Federal University of Rio de Janeiro,
Rio de Janeiro, Brazil*

INTRODUCTION

Sarcoplasmic reticulum vesicles (SRV) isolated from skeletal muscle and suspended in a medium containing Mg, ATP, and Ca can reduce the Ca concentration of the medium to less than 10^{-7} M, the level required to maintain living muscle fibers in a resting state. Oxalate or phosphate increases the Ca-storage capacity of the vesicles by providing a sink for the entering Ca (Hasselbach, 1964; de Meis, Hasselbach, and Machado, 1974). When one of these Ca-precipitating agents is present, the free Ca^{++} inside the vesicles remains constant as Ca is removed from the medium. With oxalate, a transmembrane Ca^{++} gradient of the order of 3,000:1 can be attained in the steady state.

Hydrolysis of ATP and the translocation of Ca into the vesicles involve a transfer of the γ-phosphate of ATP to a membrane protein (E), releasing ADP and forming an acyl-phosphoprotein (E $\sim$ P) (Eq. 1) (Makinose, 1967, 1969, 1973; Yamamoto and Tonomura, 1967, 1968; Martonosi, 1969; Inesi, Maring, Murphy, and McFarland, 1970; Panet, Pick, and Selinger, 1971; de Meis, 1972; Yamada and Tonomura, 1972*b*). This complex can transfer two Ca ions across the membrane and is subsequently hydrolyzed to release inorganic phosphate (P$_i$) (Eq. 2). The latter step is rate limiting for the overall reaction (Makinose, 1969; de Meis and de Mello, 1973).

$$ATP + E \rightleftharpoons E \sim P + ADP \qquad (1)$$
$$E \sim P \rightleftharpoons E + P_i \qquad (2)$$

The reversibility of both reactions in intact vesicles has been demonstrated by various means. We may summarize the evidence for formation

---

* Permanent address: Laboratory of Neurophysiology, Department of Neurology, College of Physicians and Surgeons of Columbia University, 630 W. 168th St., New York, New York 10032.

of a high-energy bond without the prior hydrolysis of ATP as follows:

(a) *Phosphorylation of the membrane protein by* $^{32}P_i$ in the presence of ATP and Ca (Makinose, 1972, 1973; de Meis and Masuda, 1974; Kanazawa and Boyer, 1973; Masuda and de Meis, 1973), in the presence of Ca alone (Yamada and Tonomura, 1972*a;* Makinose, 1972), or in the absence of both ATP and Ca (Masuda and de Meis, 1973; Kanazawa and Boyer, 1973).

(b) *Incorporation of* $^{32}P_i$ *into ATP* in the presence of Ca and ATP (Makinose, 1971; Racker, 1972; de Meis and Carvalho, 1974). The rate of this ATP $\rightleftharpoons$ P$_i$ exchange builds up as Ca phosphate is taken up by the vesicles and reaches a steady value when Ca accumulation attains a steady state. It appears to be the result of the two reactions simultaneously operating forward and backward (Makinose, 1971; Racker, 1972).

(c) *Net synthesis of* $(\gamma\text{-}^{32}P)$ *ATP* from ADP and $^{32}P_i$ in the presence of Mg by vesicles preloaded with Ca. In this case the formation of 1 mole of ATP is coupled with the very rapid release of 2 moles of Ca into a low-Ca medium (Barlogie, Hasselbach, and Makinose, 1971; Makinose and Hasselbach, 1971; Makinose, 1971, 1972, 1973; Yamada, Sumida, and Tonomura, 1972; Panet and Selinger, 1972).

In most of the studies cited, reversal was reduced or abolished by procedures that abolished the Ca^{++} gradient across the membrane, as might be expected if the Ca^{++} concentration gradient provided the energy for the formation of E $\sim$ P (Makinose, 1971, 1972, 1973; Makinose and Hasselbach, 1971; cf. Mitchell, 1966). However, in some conditions SRV membranes can catalyze ATP $\rightleftharpoons$ P$_i$ exchange (Racker, 1972) and E $\sim$ P formation from P$_i$ (Masuda and de Meis, 1973; Kanazawa and Boyer, 1973) in the absence of a Ca^{++} gradient. In this chapter we explore further the role of the Ca^{++} gradient in reversal of the ATPase reaction, using ATP $\rightleftharpoons$ P$_i$ exchange as an index of reversal. We also examine the relationship between ATP hydrolysis and ATP $\rightleftharpoons$ P$_i$ exchange under a variety of conditions.

Part of the work reported here has been published in more detail elsewhere (de Meis and Carvalho, 1974).

METHODS

SRV were prepared from rabbit skeletal muscle as described by de Meis and Hasselbach (1971) and were used within 3 days for measurements of Ca uptake. Leaky vesicles (Duggan and Martonosi, 1970) were prepared by 20-min incubation at room temperature in 1 mM Tris-EGTA (pH 9), followed by readjustment of the pH to 7 with Tris maleate and storage at 0°C for as long as 1 hr during use. In agreement with other reports (Inesi, Goodman, and Watanabe, 1967; Fiehn and Hasselbach, 1969; Duggan and Martonosi, 1970), leaky vesicles maintain ATPase activity but no longer accumulate Ca. Vesicles made leaky by adding 60 μM of the ionophore X-537A to the assay medium (Scarpa and Inesi, 1972; Pressman, 1973) are similarly affected. For some experiments, vesicles were solubilized (Ikemoto, Bhatnagar, and Gergely, 1971) with Triton X-100 in 20% glycerol at pH 8 (Triton: protein 2:1 w/w), and stored at 5°C for as long as 3 days after centrifugation in 4 mM $CaCl_2$.

Ca uptake by intact vesicles was measured with ^{45}Ca, using Millipore filters (de Meis, 1969, 1971). ATPase activity was assayed spectrophotometrically by the method of Fiske and SubbaRow (1925) or by measuring the release of $^{32}P_i$ from (γ-^{32}P) ATP following precipitation of the protein with 1.5 volumes TCA (10% w/v). The $^{32}P_i$ was extracted from the assay medium as phosphomolybdate, using a mixture of isobutanol and benzene according to a modification of the method of Avron (1960) (de Meis and Carvalho, 1974). ATP ⇌ P_i exchange was determined by measuring (γ-^{32}P) ATP formed from $^{32}P_i$ added to the medium in the presence of ADP. After the addition of TCA, the modified Avron method was used to separate excess $^{32}P_i$ from the (γ-^{32}P) ATP in the aqueous phase by two successive extractions, the second with carrier P_i added. In control experiments, the radioactivity remaining in the aqueous phase was identified as (^{32}P) ATP by paper chromatography, by column chromatography (Glynn and Chappell, 1964), and enzymatically (Panet and Selinger, 1972). Radioactive phosphate was obtained from the Brazilian Institute of Atomic Energy, purified by elution from a column of Dowex AG1-X10, and used to make (γ-^{32}P) ATP by a modification of the method of Glynn and Chappell (1964) (de Meis, 1972).

RESULTS AND DISCUSSION

Ca Uptake and ATP ⇌ P_i Exchange in Intact Vesicles

The formation of (γ-^{32}P) ATP from $^{32}P_i$ by vesicles incubated with Ca, Mg, and ATP reaches its maximum rate only when the full gradient has been developed and external Ca^{++} has been reduced to micromolar concentrations. This sequence of events is most clearly seen at the slower rates of Ca uptake obtained with lower concentrations of P_i. With 4 mM $^{32}P_i$ (*lowest solid curve,* Fig. 1), the 2-min lag observed before a significant amount of (γ-^{32}P) ATP is formed is the time required for removal of 75% of the Ca from the medium (*lowest dashed curve,* Fig. 1). The rate of ATP ⇌ P_i exchange rises to a maximum only when 98% of the Ca in the medium has been taken up.

The importance of the gradient in obtaining significant ATP ⇌ P_i exchange is further supported by the parallel decrease in Ca accumulation and ATP ⇌ P_i exchange that we observe when increasing concentrations of Triton X-100 are included in the assay medium (de Meis and Carvalho, 1974). These observations are in agreement with those reported by Makinose (1971), which led Makinose and Hasselbach (1971) to suggest that SRV were an efficient system for converting osmotic into chemical energy.

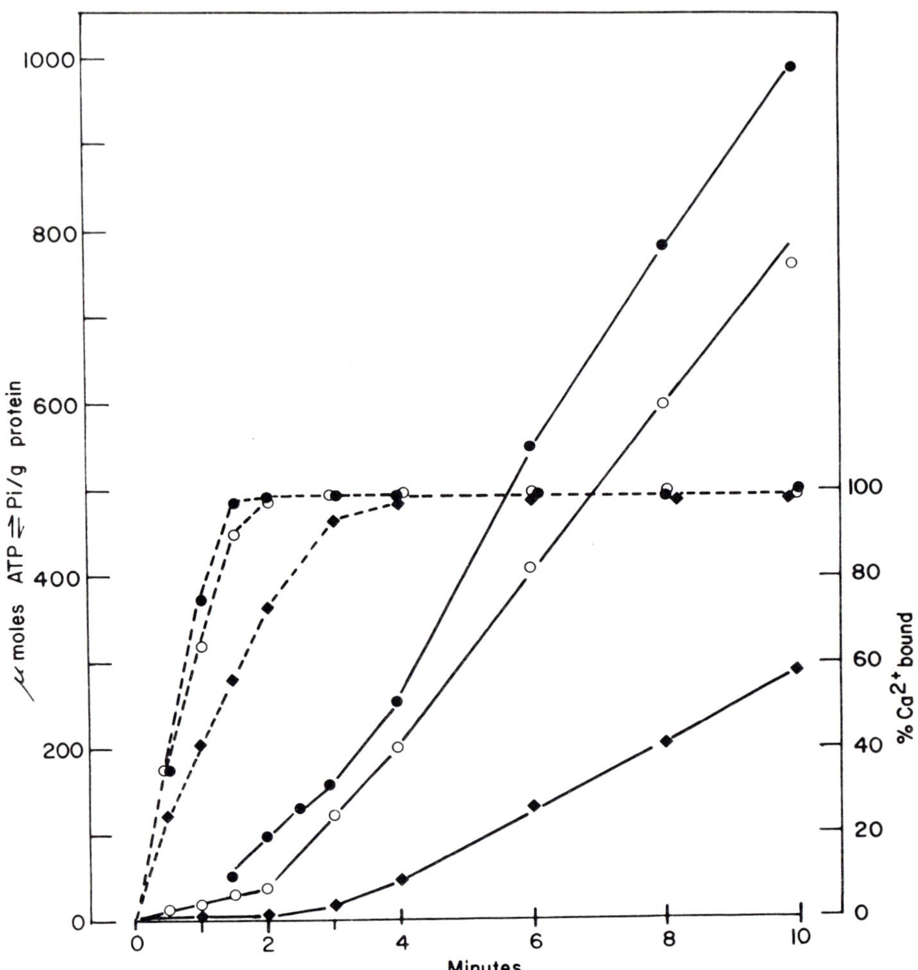

FIG. 1. Ca accumulation and rate of ATP $\rightleftharpoons$ P$_i$ exchange. Assay medium contained 4 mM ATP, 15 mM MgCl$_2$, 20 mM Tris maleate pH 6.4, 0.2 mg/ml SRV protein and 4 mM (◆), 7 mM (○) or 10 mM (●) potassium phosphate. Reaction was started by addition of SRV and stopped after different incubation intervals at 30° by removal of SRV with Millipore filters. For Ca uptake (*dashed lines*), nonradioactive P$_i$ and ^{45}Ca were used. For ATP $\rightleftharpoons$ P$_i$ exchange (*solid lines*), ^{32}P$_i$ and nonradioactive Ca were used. 0.1 mM CaCl$_2$ present in all assays.

Figure 1 also shows that the steady-state rate of ATP $\rightleftharpoons$ P$_i$ exchange in fully loaded vesicles increases with increasing P$_i$ concentrations. The apparent K_m for P$_i$, calculated from data like those in Fig. 1, is 5.6 mM (see Table 1). This value is in the same range as that reported for reversal of the Ca pump (Masuda and de Meis, 1974) under conditions that lead to net synthesis of ATP from ADP and P$_i$.

TABLE 1. *Kinetic parameters of the ATP $\rightleftharpoons$ P$_i$ exchange reaction*

Treatment	Ca (mM)	K_m (mM)	V_m (μmoles/g protein · 5 min)
A Leaky	4	39 $\pm$ 15	1240 $\pm$ 24
Intact	0.1[a]	5.6 $\pm$ 3.2	552 $\pm$ 16
B Leaky			
0 Ag	4	52 $\pm$ 32	1226 $\pm$ 30
50 μM Ag	4	7.7 $\pm$ 4.8	643 $\pm$ 60

[a] Ca outside 4–10 μM

In *A* exchange rates for leaky vesicles (0.3 mg protein/ml) were obtained from 5-min incubations at 37° and pH 6.7–6.8 in media containing from 2–10 mM ^{32}P$_i$, 10 mM ATP, 20 mM MgCl$_2$, 30 mM Tris maleate, and Ca, as indicated. For intact vesicles, rates at 5 min were calculated from the linear portions of ATP $\rightleftharpoons$ P$_i$ exchange curves like those in Fig. 1, when more than 90% of the Ca in the medium had been taken up by the vesicles. The assay media (pH 6.6, 37°) contained 2–10 mM ^{32}P$_i$, 5 mM ATP, 15 mM MgCl$_2$, 30 mM Tris maleate, 0.1 mM CaCl$_2$, and 0.3 mg/ml SRV protein. The values of K_m and V_m were calculated from linear regressions of v/s or (for intact vesicles) v/s^2 on v for 5 concentrations of P$_i$, using the means of 4 or (for leaky vesicles) 6 experiments. The errors for V_m are the standard errors; for K_m, the 90% confidence intervals.

Values in *B* were calculated as in *A* from the means of 10 paired experiments under the conditions given for Fig. 4, but with 4 mM CaCl$_2$ and 6 concentrations of P$_i$ (1–8 mM). In every case the K_m was smaller in the presence of Ag. The large errors in the K_m values in both *A* and *B* reflect the difficulty of obtaining kinetic data at substrate concentrations so far below the K_m, as well as the considerable variability among preparations.

ATP $\rightleftharpoons$ P$_i$ Exchange without a Ca^{++} Gradient

Presence of a gradient in intact vesicles coincides with the establishment of three other conditions: millimolar concentrations of Ca^{++} inside the vesicles, micromolar concentrations of Ca^{++} outside, and a low rate of ATP hydrolysis (Hasselbach, 1964). Treatment with Triton, which abolishes the gradient, also abolishes these three conditions. Since any of them might be responsible for activating reversal, we have attempted in the following experiments to distinguish the effect of the gradient from the other factors.

In the experiments of Fig. 2, steady-state ATP $\rightleftharpoons$ P$_i$ exchange and ATP hydrolysis were measured concurrently as a function of added Ca, using intact vesicles and vesicles made leaky by the pH 9–EGTA treatment. With intact vesicles, the high rates of exchange at 0.1 mM Ca and 0.3 mM Ca correspond to a large gradient and low external Ca^{++}, since the vesicles had removed more than 95% of the Ca from the medium. At 0.7 mM both the ATP $\rightleftharpoons$ P$_i$ exchange and the gradient are much diminished since the vesicles had taken up only 60% of the external Ca (leaving 0.4 mM). At the remaining points, accumulation ranged even lower. Because of the Ca phosphate

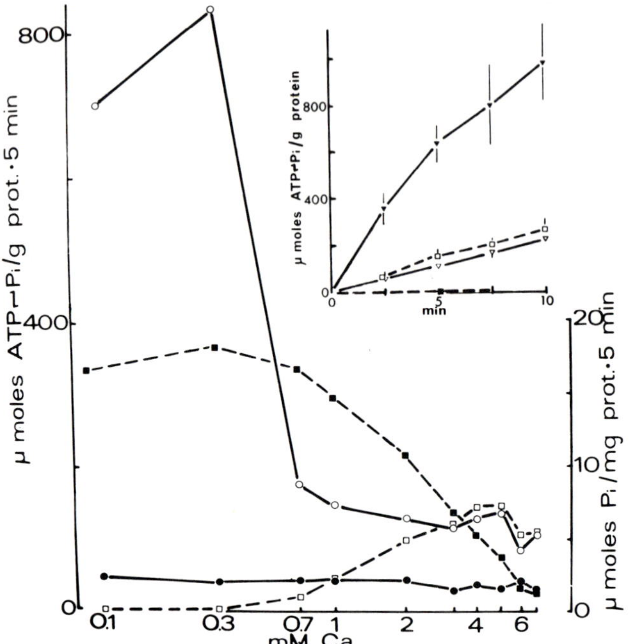

FIG. 2. Ca dependence of ATP $\rightleftharpoons$ P$_i$ exchange and ATP hydrolysis. Assay medium (37°) contained 5 or 10 mM ATP, 6 mM P$_i$, 20 mM MgCl$_2$, 30 mM Tris maleate pH 6.4, varying amounts of CaCl$_2$, and 0.3 mg/ml SRV protein. For exchange, ^{32}P$_i$ was used and the reaction was stopped with TCA (10%, w/v). In a parallel set of assays, the reaction was stopped by Millipore filtration and ATP hydrolysis was measured spectrophotometrically in the same filtrate used to determine ^{45}Ca uptake. Hydrolysis (*solid symbols*) and exchange (*open symbols*) took place in intact (*solid lines*) and leaky (*dashed lines*) SRV. Ca uptake at 5 min was nil for the leaky SRV and in intact SRV ranged from 95.7% at 0.1 mM Ca to 2.1% at 7 mM Ca (see text). ATP concentration was 5 mM for intact vesicles and 10 mM for leaky vesicles. *Inset:* ATP $\rightleftharpoons$ P$_i$ exchange at low and high Ca in intact and leaky SRV. Ten mM ATP were added to assay medium. Intact vesicles: exchange measured after 98% of Ca in the assay medium (0.1 mM) had been removed by the SRV (▼—▼); exchange measured after vesicles had been loaded in 0.1 mM Ca and the Ca concentration of the medium raised to 6 mM (▽—▽). Leaky vesicles: 0.1 mM Ca (■--■) or 6 mM Ca (□--□). Values represent the average ± SE of 3 experiments.

precipitation, internal Ca^{++} remains high throughout, at about 10 mM. Exchange thus approaches a minimum as the concentration of Ca outside the vesicles approaches the concentration inside: this minimum, however, is still a substantial fraction – 15 to 20% – of the maximum rate obtained with a gradient.

By contrast, in the experiment with leaky vesicles ATP $\rightleftharpoons$ P$_i$ exchange is practically zero at 0.1 mM Ca, but increases by several orders of magnitude as Ca is increased to 5 mM. The maximum rate in leaky vesicles is nearly identical to the rate in the intact vesicles at the same Ca concentration: in both cases the concentrations on both sides of the membrane are of

TABLE 2. *ATP* ⇌ *P$_i$ exchange and hydrolysis in SRV (μmoles/g protein · 5 min)*

Vesicle preparation	ATP ⇌ P$_i$ exchange		Hydrolysis	
	0.1 mM Ca	6–10 mM Ca	0.1 mM Ca	6–10 mM Ca
Leaky (pH 9–EGTA)	2.6 ± 1.6	125 ± 22	12.8 ± 1.7	2.6 ± 0.7
Solubilized	0	24.3 ± 2.1	20.3 ± 0.9	6.4 ± 1.4
Intact	465 ± 107[a]	89 ± 10	1.9 ± 0.5[a]	1.3 ± 0.5

[a] Ca outside: 4–10 μM

Experimental conditions as in Fig. 2, except that pH was 6.6–6.8 for solubilized vesicles and 10 mM ATP was used in all experiments. Values represent means ± SE of 5 or 6 experiments.

the same order of magnitude. A Ca concentration of about 2 mM (range 1.6 to 2.4 mM) was required for half-maximal activation of exchange in leaky vesicles and, in similar experiments, in solubilized vesicles.

Table 2 gives the results of experiments in which ATP ⇌ P$_i$ exchange and ATP hydrolysis were measured in leaky vesicles, solubilized vesicles and in intact vesicles in high Ca — three preparations in which the Ca^{++} gradient is abolished. Each preparation was tested in low Ca (0.1 mM) and in a Ca concentration (6 to 10 mM) of the order of that facing the internal surface of the vesicles when they are fully loaded in the presence of P$_i$. In three preparations, exchange and hydrolysis in 0.1 mM Ca were also measured in untreated vesicles in the presence of the ionophore X-537A at 60 μM, which completely blocks Ca accumulation. The results were similar to those shown in Table 2 for leaky vesicles in 0.1 mM Ca.

The data of Table 2 and Fig. 2 show that (a) exchange in the complete absence of a gradient can reach 20 to 25% of the rate in intact vesicles maintaining a gradient; (b) 4 to 6 mM Ca is necessary and sufficient for this effect; and (c) in loaded vesicles either the gradient or the low concentration of Ca^{++} facing the external membrane additionally activates the rate of ATP ⇌ P$_i$ exchange. The similarity in rates of exchange for leaky and intact vesicles in high Ca suggests that in this condition the kinetic parameters of the reaction with P$_i$ in the two preparations may be identical. Conversely, we will show below that the activation of exchange associated with the gradient reflects a change in the K_m for P$_i$.

Comparison of Hydrolysis and Exchange Rates

For the first two preparations in Table 2, and for the leaky vesicles in Fig. 2, a higher rate of ATP ⇌ P$_i$ exchange is associated with a lower rate of the forward reaction (ATP hydrolysis). With intact vesicles, however, the rates of exchange and hydrolysis are not reciprocally related. The low rate of hydrolysis in high Ca concentrations is almost identical to that obtained with a gradient at low Ca levels, where the rate of exchange is sixfold higher.

Activation of Exchange by Ag

A differential effect of ionic Ag on the forward and reverse reactions of the ATPase in SRV has provided us with another means of testing the relationship between hydrolysis and exchange. It has also allowed us to obtain with leaky vesicles exchange rates as high as those found in the presence of a gradient. Ag ions in the range of 5 to 100 μM inhibit oxalate- or phosphate-stimulated Ca accumulation in intact vesicles; and in leaky vesicles there is a parallel fall in the Ca-dependent ATP hydrolysis (Fig. 3). Ag was found by Walter and Hasselbach (1973) to be similarly potent in inhibiting the major ATPase activity of delipidated SR membranes.

At a Ag concentration in which the effect on Ca uptake in intact vesicles was maximal (Fig. 3), ATPase activity and ATP ⇌ P$_i$ exchange were measured in leaky vesicles at different Ca concentrations (Fig. 4). In the absence of Ag (see also Table 2 and Fig. 2), increasing Ca caused a sevenfold decrease in hydrolysis and an increase in exchange. With 50 μM Ag, exchange was also activated as Ca was increased to 4 mM, but in contrast to the control, the Ca-dependent ATP hydrolysis was completely inhibited at all

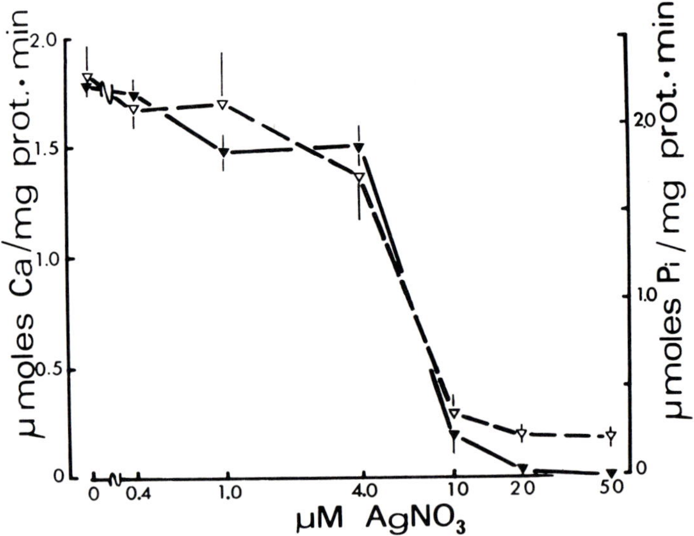

FIG. 3. Effect of Ag on Ca uptake and ATP hydrolysis. Intact (▼—▼) or leaky (▽—▽) SRV (0.1 mg/ml) were added to an assay medium containing AgNO$_3$, 5 mM ATP, 0.1 mM CaCl$_2$, 8 mM P$_i$, 20 mM MgCl$_2$, and 30 mM Tris maleate pH 6.8–7.1. The reactions were stopped after 4 min at 30° by the addition of TCA (10%, w/v) for the measurement of ^{32}P$_i$ liberated from (γ-^{32}P) ATP in leaky vesicles, or by Millipore filtration for ^{45}Ca uptake in intact sivesicles. Values represent the average ± SE of 4 experiments.

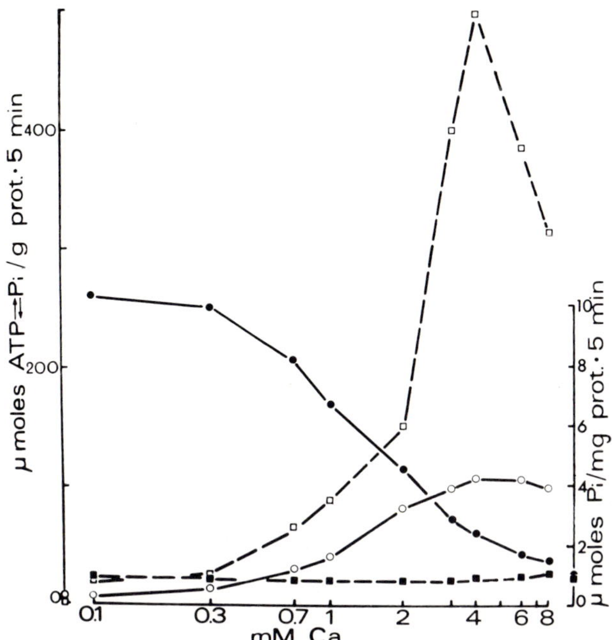

FIG. 4. Ca dependence of ATP $\rightleftharpoons$ P$_i$ exchange and ATP hydrolysis in the presence of Ag. Leaky SRV (0.3 mg/ml) were added to reaction media containing zero (*solid lines*) or 50 μM (*dashed lines*) AgNO$_3$, 5 mM ATP, 0.5 mM ADP, 6 mM P$_i$, 20 mM MgCl$_2$, 30 mM Tris maleate pH 6.8, and varying concentrations of CaCl$_2$. For hydrolysis ($\bullet$, $\blacksquare$), (γ-^{32}P) ATP was used; for exchange ($\bigcirc$, $\square$), ^{32}P$_i$ was used. The reactions were terminated at 5 min by adding TCA (10%, w/v). Rates found in the absence of Ca (1.5 mM EGTA) are indicated on the two ordinates.

Ca concentrations. The maximal rate of exchange in the presence of Ag was four to five times that obtained in the control.

Thus in high Ca — and therefore without a gradient and without the low external Ca^{++} levels that accompany the establishment of a gradient — we can obtain incorporation of P$_i$ into ATP at rates that are nearly as high as those found in the presence of a gradient. Furthermore, the activation of this exchange with increasing Ca concentrations is not associated with any change in the hydrolysis rate.

Kinetic Parameters of ATP $\rightleftharpoons$ P$_i$ Exchange with and without a Ca^{++} Gradient

The foregoing experiments demonstrate that a Ca^{++} concentration gradient is not necessary for significant reversal of phosphorylation as long as the enzyme is exposed to high Ca concentrations, and, further, that the activation of reversal is not correlated with any particular effect on the rate of the forward reaction. Previous experiments have demonstrated that the nucleotide and P$_i$ phosphorylate the same site (de Meis and Masuda, 1974); in loaded vesicles the affinity for ATP is about 10^5 times that for P$_i$ (de Meis and de Mello, 1973; de Meis and Masuda, 1974; and Fig. 1). Taken together, these observations suggest that activation of reversal reflects both a Ca requirement and a Ca-regulated competition between ATP and P$_i$ or ADP for the phosphorylation site. We might predict that conditions that favor reversal would lower the K_m for ADP or P$_i$.

The requirement for ADP in the ATP $\rightleftharpoons$ P$_i$ exchange reaction is very low and we have not been able to measure the K_m (de Meis and de Mello, 1973; de Meis and Carvalho, 1974).

For P$_i$, however, it is possible to determine the K_m in intact and in leaky vesicles at the Ca concentrations that maximally activate each type of preparation (see Fig. 2 and Table 2). The affinity for P$_i$ (Table 1,*A*) is nearly 10 times greater in the presence of a gradient than in leaky vesicles. In four experiments with solubilized vesicles in 4 mM Ca, the K_m was also larger than in intact vesicles, but the variability among preparations was quite large. V_m is increased by the pH 9–EGTA treatment, but falls to much lower values when the enzyme is solubilized.

The activation of exchange by Ag in leaky vesicles at optimal Ca concentrations is also associated with a decrease in the K_m for P$_i$ (Table 2,*B*). With this treatment, the V_m is restored to the level of that in intact vesicles. We are currently investigating the mechanism of these effects of Ag. It is clear, however, that Ag permits us to imitate the effect of the Ca^{++} gradient in greatly reducing the K_m for P$_i$, thus activating the ATP $\rightleftharpoons$ P$_i$ exchange reaction.

Ca Regulation of Hydrolysis and ATP $\rightleftharpoons$ P$_i$ Exchange

The increased affinity for P$_i$ observed in the presence of a gradient may be a consequence of the low external Ca^{++} concentration rather than the gradient itself. The degree of membrane phosphorylation by ATP varies with the Ca^{++} concentration of the medium (Yamamoto and Tonomura, 1967; Makinose, 1969; de Meis, 1972; de Meis and de Mello, 1973); and at low external Ca^{++} levels, the fraction of membrane sites phosphorylated by ATP is less than one half (de Meis, 1972; de Meis and de Mello, 1973). Since the total number of phosphorylated membrane sites remains constant at different ratios of NTP and P$_i$ (de Meis and Masuda, 1974), sites not phosphorylated by ATP would be phosphorylated by P$_i$. Between 5 and 10 mM Ca, on the other hand, more than half the sites are phosphorylated by ATP, leaving fewer sites available for the reaction with P$_i$ (de Meis, 1972; de Meis and de Mello, 1973). With 0.1 mM Ca on both sides of the membrane, all the membrane sites will be phosphorylated by ATP (de Meis and de Mello, 1973; de Meis and Masuda, 1974); accordingly, no ATP $\rightleftharpoons$ P$_i$ exchange can be measured (Figs. 2 and 4, and Table 2, *leaky vesicles*).

If this Ca-regulated competition between ATP and P_i does not always lead to reciprocity of exchange and hydrolysis rates, it may be because hydrolysis, measured as the release of P_i, can be inhibited either in the first or in the second step of the overall reaction described in the introduction. Only inhibition of the first step (E ~ P formation from P_i) would facilitate phosphorylation by P_i. Such an inhibition would be consistent with both the low hydrolysis rate and the decrease in K_m that accompany activation of ATP ⇌ P_i exchange in leaky vesicles at high Ca concentrations in the presence of Ag.

It is implicit in the data (Table 1,A) comparing leaky and intact vesicles that at adequate Ca levels rates of exchange as high as those obtained with a gradient would be observed in both preparations if they could be tested at P_i concentrations above the K_m. The high rates of exchange obtained with leaky vesicles when the K_m is lowered by Ag (Fig. 4) support this inference. Unfortunately, in our experimental conditions we are limited by the solubility of Ca phosphate to a P_i concentration of about 10 mM.

The experiments we have described suggest that the role of the gradient in activating reversal may not be one of providing osmotic energy for ATP formation, as has been proposed (Makinose, 1972, 1973). We cannot exclude the possibility that such an energy transfer may occur in intact vesicles, but we have clearly shown that it is not required for the ATP ⇌ P_i exchange reaction.

The possibility that the hydrolysis of ATP itself furnishes the energy for the incorporation of P_i into ATP is rendered less likely by the results reported here. They suggest that ATP hydrolysis and ATP formation from P_i are competitive reactions.

ACKNOWLEDGMENTS

M. da G. C. C. and M. M. S. hold fellowships from the Brazilian National Research Council (CNPq). This investigation was also supported in part by the Council for Graduate Education of the Federal University of Rio de Janeiro and the Brazilian National Bank for Economic Development (FUNTEC 241).

The antibiotic X-537A was the kind gift of Dr. Julius Berger of Hoffman-La Roche, Inc.

REFERENCES

Avron, M. (1960). Photophosphorylation by Swiss-chard chloroplasts. *Biochim. Biophys. Acta* 40:257–272.

Barlogie, B., Hasselbach, W., and Makinose, M. (1971): Activation of calcium efflux by ADP and inorganic phosphate. *FEBS Lett.* 12:267–268.

de Meis, L. (1969): Ca^{++} uptake and acetyl phosphatase of skeletal muscle microsomes. *J. Biol. Chem.* 244:3733–3739.

de Meis, L. (1971): Allosteric inhibition by alkali ions of the Ca^{++} uptake and adenosine triphosphatase activity of skeletal muscle microsomes. *J. Biol. Chem.* 246:4764–4773.

de Meis, L. (1972): Phosphorylation of the membrane protein of the sarcoplasmic reticulum. Inhibition by Na^+ and K^+. *Biochemistry* 11:2460–2465.

de Meis, L., and Carvalho, M. G. C. (1974): The role of the Ca^{++} concentration gradient in the $ATP \rightleftharpoons P_i$ exchange reaction catalyzed by sarcoplasmic reticulum. *Biochemistry.* 13:5032–5038.

de Meis, L., and de Mello, M. C. F. (1973): Substrate regulation of membrane phosphorylation and of Ca^{++} transport in sarcoplasmic reticulum. *J. Biol. Chem.* 248:3691–3701.

de Meis, L., and Hasselbach, W. (1971): Acetyl phosphate as substrate for Ca^{++} uptake in skeletal muscle microsomes. Inhibition by alkali ions. *J. Biol. Chem.* 246:4759–4763.

de Meis, L., Hasselbach, W., and Machado, R. D. (1974): Characterization of calcium oxalate and calcium phosphate deposits in sarcoplasmic reticulum vesicles. *J. Cell Biol.* 62:505–509.

de Meis, L., and Masuda, H. (1974): Phosphorylation of the sarcoplasmic reticulum membrane by orthophosphate through two different reactions. *Biochemistry* 13:2057–2062.

Duggan, P. F., and Martonosi, A. (1970): Sarcoplasmic reticulum. IX. The permeability of sarcoplasmic reticulum membranes. *J. Gen. Physiol.* 56:147–167.

Fiehn, W., and Hasselbach, W. (1969): The effect of diethylether upon the function of the vesicles of sarcoplasmic reticulum. *Eur. J. Biochem.* 9:574–578.

Fiske, C. H., and SubbaRow, Y. (1925): The colorimetric determination of phosphorus. *J. Biol. Chem.* 66:375–400.

Glynn, I. M., and Chappell, J. B. (1964): A simple method for the preparation of ^{32}P-labelled adenosine triphosphate of high specific activity. *Biochem. J.* 90:147–149.

Hasselbach, W. (1964): Relaxation and the sarcotubular calcium pump. *Fed. Proc.* 23:909–912.

Ikemoto, N., Bhatnagar, G. M., and Gergely, J. (1971): Fractionation of solubilized sarcoplasmic reticulum. *Biochem. Biophys. Res. Commun.* 44:1510–1517.

Inesi, G., Goodman, J. J., and Watanabe, S. (1967): Effect of diethyl ether on the adenosine triphosphatase activity and the calcium uptake of fragmented sarcoplasmic reticulum of rabbit skeletal muscle. *J. Biol. Chem.* 242:4637–4643.

Inesi, G., Maring, E., Murphy, A. J., and McFarland, B. H. (1970): A study of the phosphorylated intermediate of sarcoplasmic reticulum ATPase. *Arch. Biochem. Biophys.* 138:285–294.

Kanazawa, T., and Boyer, P. D. (1973): Occurrence and characteristics of a rapid exchange of phosphate oxygens catalyzed by sarcoplasmic reticulum vesicles. *J. Biol. Chem.* 248:3163–3172.

Makinose, M. (1967): Abstract. *Second Intl. Congr. Biophysics, Vienna, 1966*, p. 276.

Makinose, M. (1969): The phosphorylation of the membrane protein of the sarcoplasmic vesicles during active calcium transport. *Eur. J. Biochem.* 10:74–82.

Makinose, M. (1971): Calcium efflux dependent formation of ATP from ADP and orthophosphate by the membranes of the sarcoplasmic vesicles. *FEBS Lett.* 12:269–270.

Makinose, M. (1972): Phosphoprotein formation during osmo-chemical energy conversion in the membrane of the sarcoplasmic reticulum. *FEBS Lett.* 25:113–115.

Makinose, M. (1973): Possible functional states of the enzyme of the sarcoplasmic calcium pump. *FEBS Lett.* 37:140–143.

Makinose, M., and Hasselbach, W. (1971): ATP synthesis by the reverse of the sarcoplasmic calcium pump. *FEBS Lett.* 12:271–272.

Martonosi, A. (1969): Sarcoplasmic reticulum. VII. Properties of a phosphoprotein intermediate implicated in calcium transport. *J. Biol. Chem.* 244:613–620.

Masuda, H., and de Meis, L. (1973): Phosphorylation of the sarcoplasmic reticulum membrane by orthophosphate. Inhibition by calcium ions. *Biochemistry* 12:4581–4585.

Masuda, H., and de Meis, L. (1974): Calcium efflux from sarcoplasmic reticulum vesicles. *Biochim. Biophys. Acta* 332:313–315.

Mitchell, P. (1966): Chemiosmotic coupling in oxidative and photosynthetic phosphorylation. *Biol. Rev.* 41:445–502.

Panet, R., Pick, U., and Selinger, Z. (1971): The role of calcium and magnesium in the adenosine triphosphatase reaction of sarcoplasmic reticulum. *J. Biol. Chem.* 246:7349–7356.

Panet, R., and Selinger, Z. (1972): Synthesis of ATP coupled to Ca^{++} release from sarcoplasmic reticulum vesicles. *Biochim. Biophys. Acta* 255:34–42.

Pressman, B. (1973): Properties of ionophores with broad range cation selectivity. *Fed. Proc.* 32:1698–1703.

Racker, E. (1972): Reconstitution of a calcium pump with phospholipids and a purified Ca^{++}-adenosine triphosphatase from sarcoplasmic reticulum. *J. Biol. Chem.* 247:8198–8200.

Scarpa, A., and Inesi, G. (1972): Ionophore mediated equilibration of calcium ion gradients in fragmented sarcoplasmic reticulum. *FEBS Lett.* 22:273–276.

Walter, H., and Hasselbach, W. (1973): Properties of the calcium-dependent ATPase of the membranes of the sarcoplasmic reticulum delipidated by the nonionic detergent Triton X-100. *Eur. J. Biochem.* 36:110–119.

Yamada, S., Sumida, M., and Tonomura, Y. (1972): Reaction mechanism of the Ca^{++}-dependent ATPase of sarcoplasmic reticulum from skeletal muscle. VIII. *J. Biochem.* 72:1537–1548.

Yamada, S., and Tonomura, Y. (1972*a*): Phosphorylation of the Ca^{++}-Mg^{++} dependent ATPase of the sarcoplasmic reticulum coupled with cation translocation. *J. Biochem.* 71:1101–1104.

Yamada, S., and Tonomura, Y. (1972*b*): Reaction mechanism of the Ca^{++}-dependent ATPase of sarcoplasmic reticulum from skeletal muscle. VII. *J. Biochem.* 72:417–425.

Yamamoto, T., and Tonomura, Y. (1967): Reaction of the Ca^{++}-dependent ATPase of sarcoplasmic reticulum from skeletal muscle. I. *J. Biochem.* 62:558–575.

Yamamoto, T., and Tonomura, Y. (1968): Reaction mechanism of the Ca^{++}-dependent ATPase of sarcoplasmic reticulum from skeletal muscle. II. *J. Biochem.* 64:137–145.

Concepts of Membranes in Regulation and Excitation, edited by M. Rocha e Silva and G. Suarez-Kurtz. Raven Press, New York © 1975.

Ca Regulation in "Skinned" Muscle Fibers

John P. Reuben, Donald S. Wood, Jenny R. Zollman, and Philip W. Brandt

H. Houston Merritt Clinical Research Center for Muscular Dystrophy and Related Diseases, Departments of Neurology and Anatomy, College of Physicians and Surgeons, Columbia University, New York, New York 10032

INTRODUCTION

The sarcolemma of single muscle fibers can be removed mechanically or disrupted by chemical means to allow experimental control of the diffusible constituents within the fiber. Such "skinned" fibers are a convenient preparation for exploring processes involved in the regulation of muscle activity. One process, the major focus of this chapter, is the accumulation and release of Ca from the sarcoplasmic reticulum (SR) which, along with remnants of the transverse tubular system (TTS), remains after skinning (Ebashi and Endo, 1968).

Mechanical removal of sarcolemma from single fibers was first described for frog muscle by Natori (1954) and later by Endo, Tanaka, and Ogawa (1970) in Japan and Podolsky and his collaborators in the United States (Constantin and Podolsky, 1967). In the original technique, the fibers were placed in paraffin oil during the mechanical skinning procedure and then dipped into different experimental solutions. The surface membranes of crayfish and frog fibers can be removed, however, in an aqueous solution containing a calcium chelator: [ethylene glycol-bis-(β-aminoethyl ether) N, N'-tetraacetic acid (EGTA) or ethylenediaminetetraacetic acid (EDTA)], Mg, adenosine triphosphate (ATP), and an appropriate salt (Reuben, Brandt, and Grundfest, 1967; Reuben, Brandt, Berman, and Grundfest, 1971).

In some muscles, such as heart (Winegrad, 1971) or human skeletal muscle, mechanical removal of the sarcolemma is not necessary. The surface membranes of those fibers become porous to ions of molecular weights up to at least 550 (e.g., ATP) when exposed to aqueous solutions containing millimolar concentrations of either EGTA or EDTA. The chemical skinning procedure is of particular value in studies of human skeletal muscle because the porous characteristics of the fiber persist after removal of the EGTA or EDTA and subsequent exposure to high concentrations of ionized Ca (Reuben, Brandt, Sorenson, and Eastwood, 1974*b*). In contrast, chemically skinned heart muscle loses its permeability to high molecular weight ions on addition of excess ionized Ca (Winegrad, 1971). Crayfish skeletal muscle fibers do not respond to chemical treatment and require mechanical skinning—implying fundamental differences in the sarcolemmal membranes of these muscles.

The physiological state of the internal membrane system in skinned fibers can be evaluated by measuring the accumulation and release of Ca. In practice, the Ca-regulating properties of skinned fibers are analyzed by following the time course and amplitude of isometric tensions, by observing opacity changes of fibers when Ca oxalate precipitates within matrix compartments during net Ca uptake, and by chemical determination of total Ca.

In the present study, isometric tension was the primary measure of change in the distribution of Ca within the filament matrix; therefore the effects of the other critical solution components on the contractile proteins had to be evaluated. A formulation for this evaluation (Reuben et al., 1971; Brandt, Reuben, and Grundfest, 1972), based on experiments on crayfish fibers, describes quantitatively the dependence of tension in the absence of Ca regulation on the concentration of ATP, Mg, MgATP, and Ca. This formulation does not provide a molecular interpretation of actin-myosin interaction, but accurately predicts the degree of contractile activation in fibers exposed to a wide range of concentrations of the critical ions, Ca and MgATP. With this background, the amplitude and time course of isometric tensions can be used to study the Ca-regulating systems.

The data generated from studies using skinned-fiber preparations has added a great deal to our understanding of Ca regulation of muscle, but the picture is still far from complete. A melding of insights gained from studies using isolated SR (see de Meis, *this volume*) with those based on skinned-fiber research is necessary. For this reason we frequently refer to work on extracted SR throughout the following report.

EXPERIMENTAL APPROACH

Skinning Solution

The solution in which fibers are skinned is called a relaxing (R) solution. The basic composition is the same for both human and crayfish preparations. The constituents of solution R are: 140 mM K propionate (for human fibers), 185 mM K propionate (for crayfish fibers); 5 to 10 mM EGTA; 1 to 2 mM Mg; 2 to 5 mM ATP; 5 mM imidizole (pH 7.00). Other K salts may be used, but problems arise if the anion of the salt is permeant. For example, in the crayfish a presoak is necessary prior to mechanical skinning, and a permeant ion such as Cl makes the fiber swell. The rationale for the composition of solution R is straightforward: the K concentration should be similar to that in the myoplasm of intact fibers. The EGTA is necessary to keep ionized Ca below levels that initiate contractile activation. The MgATP maintains actin and myosin in a dissociated state. Apparently only the reduction of ionized Ca in the bath by EGTA is necessary for the chemical skinning of human skeletal muscle fibers.

Experimental Chamber

The experimental chamber (Fig. 1) provides temperature regulation, vigorous and continuous mixing, a vacuum system for changing solutions, and adjustable clamps for holding single muscle fibers. One clamp is attached to a strain gauge and is used to adjust sarcomere spacings so that maximum isometric tension is attained during activation. An important feature is the stirring apparatus. The attainment of a given concentration of reactants within the filament matrix depends on diffusional rates of the ions within the bath and within the matrix. Mixing must be vigorous enough to minimize unstirred layers and to ensure rapid and uniform distribution of the critical reactants at the surface of the fiber.

The Muscle Fibers

The isolated crayfish fiber (see Brandt, Reuben, Girardier, and Grundfest, 1965) is about 5 mm in length, with an average diameter of 200 μ. Some sarcolemma is left at each end of the fiber to serve as a protective cover for the myofibrils when the cut ends are clamped (Fig. 1A). The human fibers (Fig. 1B), mainly from leg muscle (gastrocnemius or quadriceps) biopsies, are approximately 70 μ in diameter and cut to about 4 mm in length. The human fibers are durable when stored at 5°C in solution R; maximal tension can be elicited after 2 weeks, and the Ca-regulating properties of the internal membranes remain for about 5 days.

EVIDENCE OF Ca SEQUESTRATION IN THE SKINNED FIBER

It has been demonstrated many times with a variety of techniques that skinned fibers are capable of net Ca uptake (see, for example, Ford and Podolsky, 1972; Orentlicher, Reuben, Grundfest, and Brandt, 1974). Of particular interest are biochemical studies of the isolated SR in which oxalate is used to enhance the net uptake of Ca (see Hasselbach and Makinose, 1961, 1963; Hasselbach, Makinose, and Fiehn, 1974). In the presence of oxalate, the skinned fiber darkens as Ca oxalate precipitates within intracellular spaces (Fig. 2). Moreover, Ca oxalate seems to precipitate only in compartments of the skinned fiber where net Ca uptake occurs.

The bathing solution for oxalate experiments contains a buffered concentration of ionized Ca (pCa 6.3 to 7.0) too low to activate the contractile apparatus and induce tension. The product of the concentrations of oxalate

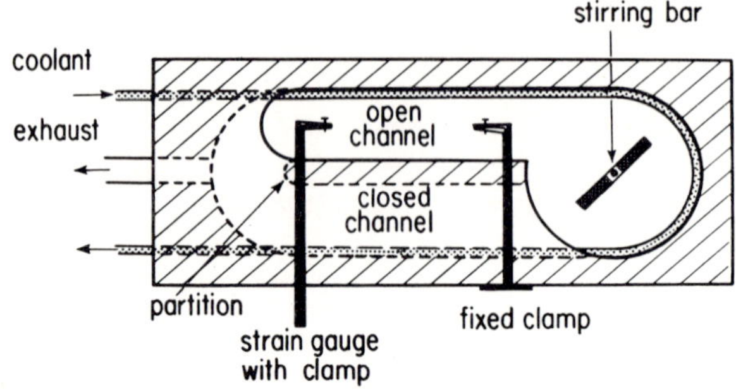

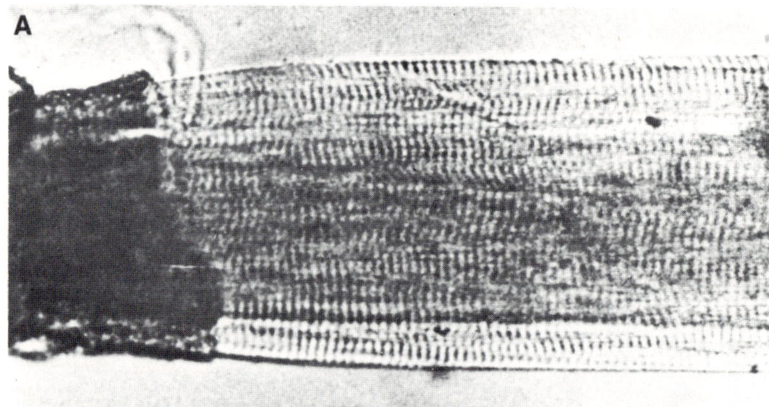

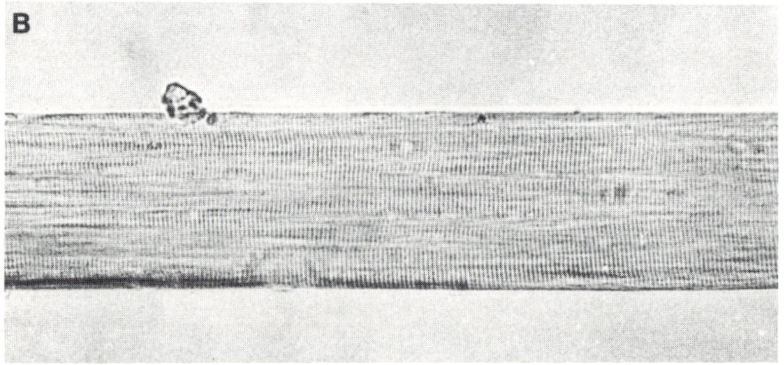

FIG. 1. Experimental chamber and muscle preparations. The skinned fiber is held between two plastic clamps in the open channel. Circulation of saline through the channels is controlled by a motor-driven stirring bar. Bath temperature is regulated by temperature-controlled water circulating through stainless steel tubing at the outer edges of the channels. The mechanically skinned crayfish fiber (A) is about 200 μ in diameter; sarcomere length is about 8 μ. The sarcolemma peeled back from most of the fiber can be seen as a dense collar at the extreme left. The chemically skinned human fiber (B) is about 70 μ in diameter; sarcomere length is approximately 3.1 μ.

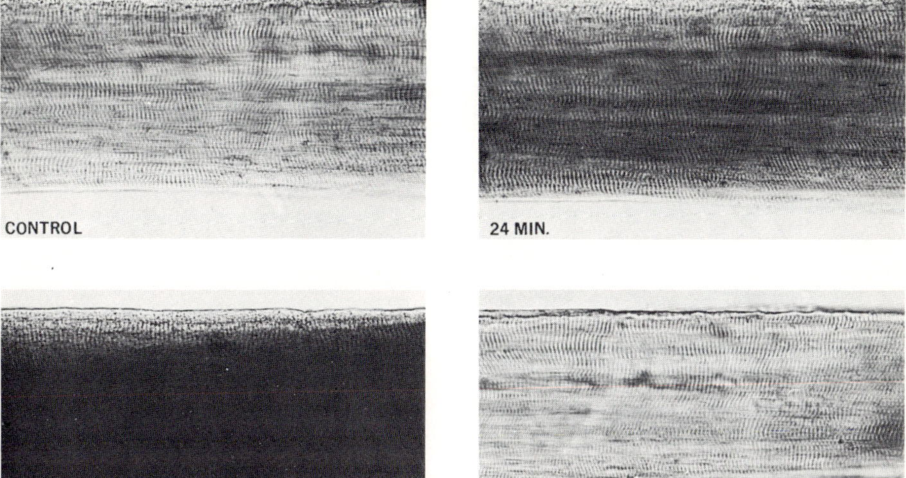

FIG. 2. Darkening of the skinned fiber during accumulation of Ca oxalate. *Top left:* fiber before the addition of buffered Ca (pCa 6.7) and oxalate (3 mM) to the bathing solution. *Top right:* Darkening after 24 min. *Bottom left:* Fiber became nearly opaque after 62 min. *Bottom right:* Addition of Brij-58 (0.5%) caused a maximum transient tension and then the fiber became transparent. The fiber remained transparent despite continued exposure to the Ca oxalate solution for several hours. (From Orentlicher et al., 1974.)

and ionized Ca within the bath is kept below the solubility product (K_s) for Ca oxalate. Consequently, unless at least one of these ions is concentrated within a compartment accessible to both, no Ca oxalate precipitate should form. Electron microscopic studies have localized Ca precipitates, such as Ca oxalate, within the SR in skinned fibers (Costantin, Franzini-Armstrong, and Podolsky, 1965; Orentlicher et al., 1974), in intact fibers (Reuben, Brandt, and Grundfest, 1974a), and within the vesicles of extracted SR (de Meis, Hasselbach, and Machado, 1974).

Further evidence that fiber darkening due to Ca precipitation involves net uptake of Ca comes from chemical analysis. After 1 hr in a buffered Ca medium (pCa 6.7) containing, for example, 3 mM oxalate, crayfish fibers gained 40 to 50 mM Ca/kg wet weight as measured fluorometrically (*unpublished observations*). The accumulated Ca can be released (inducing a tension) by treating the fiber with nonionic detergents (Brij-58, Triton X-100) that suggests the sequestered Ca was within a membrane-bound compartment (Orentlicher et al., 1974). After exposure to detergent, the fiber again appears transparent (*bottom right,* Fig. 2).

Other evidence that skinned fibers take up net quantities of Ca comes from experiments such as that shown in Fig. 3. The crayfish fiber in that experi-

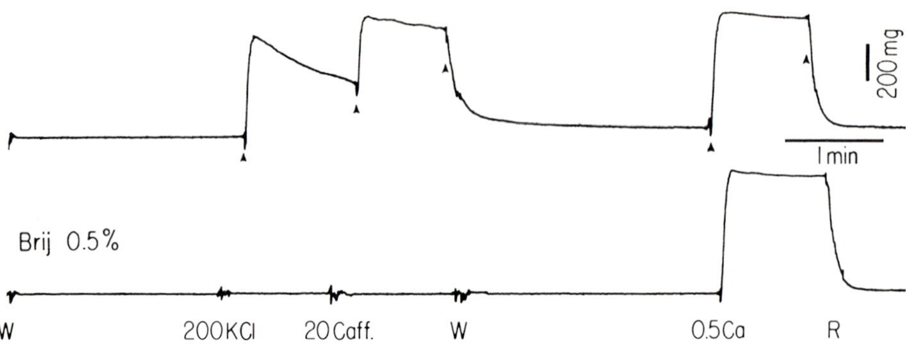

FIG. 3. Selective effects of Brij-58. *Top row:* First tension was induced by substituting KCl for the K propionate of the solution bathing the skinned crayfish fiber. Before attainment of complete relaxation, 20 mM caffeine was added to the saline, eliciting a maximum tension. The fiber was relaxed by washing out the caffeine with control propionate saline (W). The subsequent maximum tension was evoked by adding Ca (0.5 mM). *Bottom row:* Same sequence of solution changes was repeated after the fiber was exposed to 0.5% Brij for about 30 min. The fiber did not respond to either KCl or to caffeine, but 0.5 mM Ca, as before, evoked a maximum tension. Similar results were obtained in chemically skinned human fibers. (From Orentlicher et al., 1974.)

ment was first bathed in a solution containing unbuffered ionized Ca (10 μM) for about 10 min. The fiber did not contract in the solution, although the ionized Ca concentration in the bath would have caused a maximal tension if it had been buffered adequately by EGTA (see Orentlicher et al., 1974). A portion of the accumulated Ca became available for binding with troponin (Fig. 3, *first tension*) upon substitution of Cl for propionate. An additional fraction was released upon exposing the fiber to 20 mM caffeine (Fig. 3, *second tension*). Both the Cl- and caffeine-induced tensions can be abolished by free EGTA, and neither agent induces tension without the initial exposure to Ca (Hasselbach, 1964; Reuben et al., 1967, and *unpublished observations*). The induction of a transient tension by either Cl or caffeine must, therefore, result from the release and subsequent uptake of Ca by a compartment that accumulates Ca during the initial exposure.

The question of what compartment in the skinned fiber releases Ca on exposure to Cl or caffeine has never been answered adequately; both the SR and some of the TTS remain after skinning. The biochemical data on isolated SR are not conclusive either, because such preparations contain unknown amounts of non-SR membranes (Weber, 1968; Weber and Herz, 1968). Consequently, in discussing possible sites of Ca uptake or release in response to agents such as Cl or caffeine, the term "internal membrane system" is as precise as the available skinned-fiber data allow. In some studies, such as electron microscopic localization of Ca oxalate precipitates in the SR, assignment of a specific membrane-bound Ca compartment seems straightforward, as does the action of detergents in releasing Ca from this compartment. But whether the release of Ca by Cl, for example, occurs through direct effects of the anion on transverse tubules, SR, or both, remains to be answered.

EFFECTS OF Ca SEQUESTRATION ON TENSION DEVELOPMENT

If skinned fibers are incubated long enough in a solution containing Brij-58, they do not darken in the presence of oxalate or contract in response to Cl or caffeine (Fig. 3, *bottom tracings*). The response to Ca (Fig. 3), however, is unchanged after exposure to Brij. Consequently, the Brij-treated fiber (Fig. 4) that does not accumulate Ca can be used to evaluate the kinetics and amplitudes of Ca tensions elicited in fibers that do accumulate Ca.

In the experiment of Fig. 4, seven tensions were elicited in a skinned crayfish fiber; four prior to and three following exposure to 0.5% Brij-58 for 20 min. The tensions were evoked by adding Ca to a K propionate medium (1.6 mM MgATP, pH 7.00); total concentration of bath Ca was 30 μM. The ionized Ca level was lower than that due to chelation by ATP, but was still sufficient to induce maximal tension. In the first record of Fig. 4 (*top left*), addition of Ca caused a slowly rising tension with a half-time to peak amplitude of 106 sec. After relaxation in solution R (*top right*),

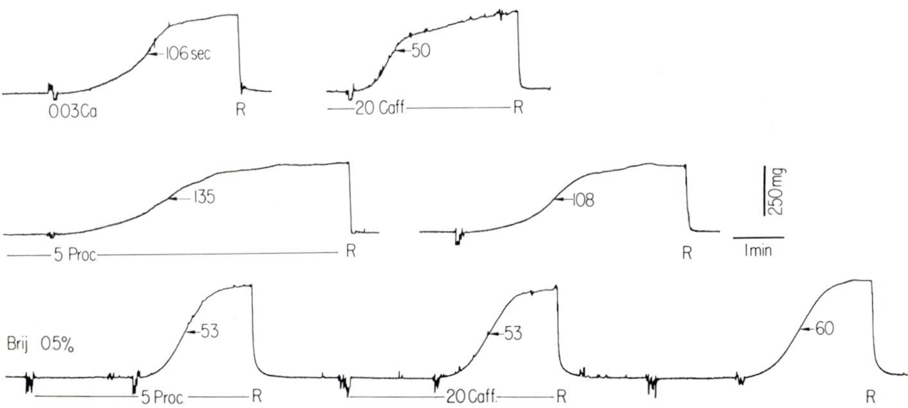

FIG. 4. Time course of tension development in a skinned crayfish fiber under conditions that modify Ca uptake. Tension was evoked in each case by adding 30 μM Ca to saline solutions containing 2 mM ATP and 2 mM Mg. Artifacts on the traces indicate the solution changes; those just preceding the tensions represent the Ca additions. *Top left:* Control tension. *Top right:* Ca-induced tension recorded while the fiber was exposed to a background of caffeine (20 mM). *Middle left:* Tension elicited after treating the fiber with procaine (5 mM). *Middle right:* A second control tension. Brij-58 was then applied for about 30 min before the continuous recording of the bottom row was obtained. The kinetics of the Ca-induced tensions were nearly identical after Brij treatment in control K propionate, in the presence of procaine, and in the presence of caffeine. (From Orentlicher et al., 1974.)

the fiber was returned to the original K propionate solution and 20 mM caffeine added. A transient tension (*not shown*) occurred upon the addition of caffeine. When the total Ca concentration in the bath was increased to 30 μM in the presence of caffeine, the fiber again contracted to the same amplitude, but the half-time of the rise to tension decreased to 50 sec. In biochemical studies, caffeine reduces the Ca-loading capacity of isolated SR (Weber and Herz, 1968). Therefore, in a skinned fiber equilibrated in caffeine, the time for Ca to attain equilibrium between bath and matrix should be reduced.

The effect on the development of isometric tension by increasing, rather than decreasing, the Ca-loading capacity of the SR was seen in the third tension (*middle left*) when the fiber was exposed to procaine (5 mM). Biochemical studies show that procaine (3.3 mM) attenuates the outflux of Ca from isolated SR without affecting the rate of Ca uptake (Weber and Herz, 1968). In the skinned fiber, agents such as procaine, which increase Ca-accumulating capacity, should increase the time for Ca equilibration between bath and matrix. The time to Ca saturation of troponin would increase and, as seen in Fig. 4, development of tension should be slower. The fourth tension (*middle right*) was elicited under control conditions, with essentially the same amplitude and kinetics as the first tension.

After treating the fiber with Brij-58 (0.5% for 30 min) the entire procedure was repeated (*bottom tracings*). The Ca tensions developed more rapidly after the Brij exposure, and procaine and caffeine had no effect on the rise time. The amplitudes of tensions before and after Brij treatment were comparable. Therefore, agents that increase (procaine), reduce (caffeine), or abolish (Brij-58) Ca accumulation by the SR exert enormous effects on the kinetics of tension without directly interacting with the contractile proteins.

Even in a Brij-treated fiber, however, in which the SR cannot retain Ca, the rate of tension development was slower than predicted by unhindered Ca diffusion within a cylinder. This slow tension development has been ascribed to the interaction of diffusible Ca with troponin and other Ca-binding sites (X).[1] The effect of these binding sites on tension kinetics can be demonstrated by presoaking Brij-treated fibers in subthreshold ionized Ca (Fig. 5), then observing the onset and rate of rise of tension in response to subsequent exposure to Ca that gives maximal tension. This presoak decreased the time for onset of tension and increased the rate at which tension developed, suggesting that during the presoak period a significant fraction of the binding sites were saturated with Ca, and that during the subsequent superthreshold Ca exposure the effective diffusion rate of Ca within the fiber matrix was enhanced.

The location of the X sites in skinned fibers is not known (Orentlicher

---

[1] The term "X sites" is used to denote all non-troponin sites within the fiber that reversibly bind Ca. The designation X reflects the fact that neither the nature nor the location of the sites is known.

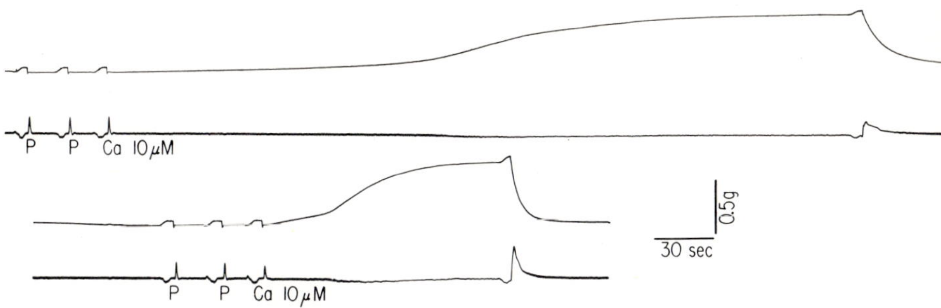

FIG. 5. Kinetics of tension development in the absence of Ca sequestration by the internal membrane system. *Top line:* Brij-treated fiber had been equilibrated in a solution buffered for pCa 8. It was then washed twice with propionate saline (P) at 20-sec intervals and 20 sec later exposed to 10 μM Ca. The tension rose slowly to a steady level before it was terminated by washing the fiber with a subthreshold Ca-buffered solution (pCa 6.2). *Bottom line:* After 2 min in the pCa 6.2 solution, the fiber was again washed twice with propionate saline and then exposed to 10 μM Ca. Tension rose more rapidly to the same steady level as attained previously. The fiber was relaxed by applying a pCa 8-buffered solution. (From Orentlicher et al., 1974.)

et al., 1974). Some may be associated with SR membranes (Fiehn and Hasselbach, 1970; Carvalho, 1968). Others may be concentrated near the H zone (Davis, Matthews, and Martin, 1974). Previous explanations of tension kinetics of skinned fibers ascribed the slowness entirely to Ca sequestration by the SR (Hellam and Podolsky, 1969). Since non-troponin Ca-binding sites affect the kinetics of tension development in skinned fibers, they must have a role in regulating tension. We will return to this point in a later section.

INDUCING TENSION BY PERTURBING THE Ca-SEQUESTERING SYSTEM

General Considerations

Tensions induced by anion substitution or addition of caffeine rise more rapidly than tensions induced by adding ionized Ca to the bath (Fig. 3). This comparison makes obvious a fact worth stressing. Intracellular compartments are relatively efficient in rapidly delivering Ca to troponin sites in response to an appropriate stimulus such as anion substitution or addition of caffeine. But the tensions induced by mobilizing Ca stored in compartments within the skinned fibers develop still more slowly than those in intact fibers in response, for example, to K depolarization. This slow time course is caused, in part, by the time required for Cl and caffeine to diffuse to the fiber's core in amounts sufficient to initiate release of Ca from the internal membrane system. Relaxed skinned fibers that have been preexposed to Ca respond with a transient tension to various changes in the composition

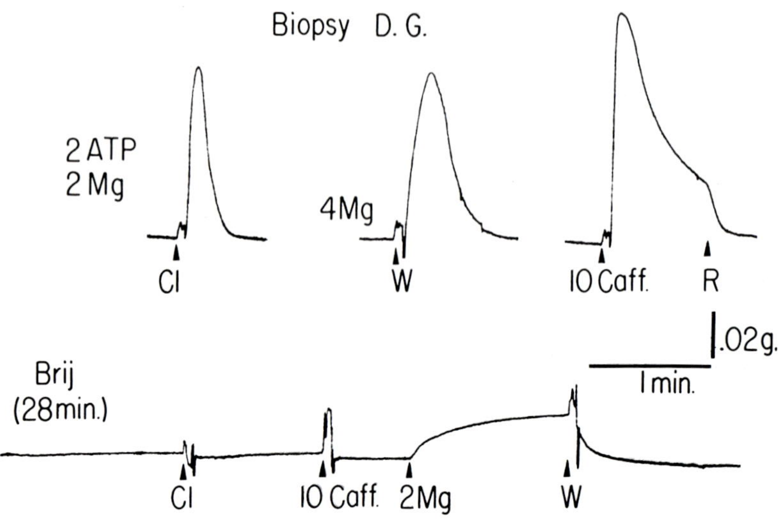

FIG. 6. Transient tensions resulting from the release and reuptake of stored Ca. *Top row of tracings:* The 1st tension (*left*) was elicited by first exposing the human skinned fiber to a low level of Ca, and then replacing the K propionate medium with a KCl medium. After preloading the fiber with Ca in the presence of 4 mM Mg, the 2nd tension (*middle*) followed a reduction in the Mg concentration to the control level of 2 mM. The 3rd tension was elicited by applying caffeine (10 mM) after the standard Ca preloading procedure. *Bottom tracings:* After treatment with Brij-58 (0.5%) for 28 min, the fiber no longer responded to Cl or caffeine. However, with addition of 2 mM Mg, the tension increased. It returned to resting level upon withdrawal of the excess Mg. Further description in the text.

of the bathing medium. Figure 6 shows conditions typical of those that produce transient tensions in human and crayfish skinned fibers. In the first record, tension was elicited by substituting Cl for propionate. In the second, the tension rose in response to a reduction in bath Mg (MgATP concentration kept constant), whereas the third tension was induced by addition of caffeine. Each of the transient tensions requires a functional system for Ca uptake, as demonstrated by treating the fibers with Brij-58. Brij treatment abolishes the transient tensions, but a small change in tension follows addition and removal of Mg (Fig. 5, *bottom tracing*). This effect of Mg on Brij-treated fibers is discussed later.

Before Brij treatment, however, anion substitution or caffeine addition or Mg withdrawal leads to the same result—transient tension— although each agent apparently induces the release of Ca in a unique manner. How perturbation of the Ca-sequestering system causes sufficient elevation of ionized Ca to cause tension is considered in the following section.

Effect of Anions

In skinned fibers capable of active Ca uptake, replacing propionate with different anions induces transient tension. The effectiveness of these anions in eliciting tension follows the order: $SCN > I > Br > NO_3 > Cl$ (Reuben et al., 1967). Furthermore, the amplitude of tension increases as larger fractions of a given anion are substituted for propionate. The mechanism by which anion substitution causes transient tension is the subject of considerable speculation. One hypothesis assumes that some portion of the internal membrane system maintains an electrical transmembrane potential, the magnitude of which depends on the anion composition within the fiber. Proponents of this hypothesis (Costantin and Podolsky, 1967) suggest that replacing propionate with an anion such as Cl causes a depolarization that triggers Ca release. Unfortunately, the physical dimensions of the internal membrane system (tubules and SR) do not allow a critical electrophysiological test.

Our own speculations are based on the observed order of anion effectiveness in mobilizing Ca (inducing tension) within skinned fibers. The absorbability of anions to membranes frequently follows the same anion sequence (see Hober, 1945; Sollner, 1949). Physiological functions that change according to this order include reduction of mechanical threshold in frog fibers (Hodgkin and Horowicz, 1960) and increased sensitivity of crayfish fibers to injected Ca (Reuben, Brandt, Katz, and Grundfest, 1970; Reuben et al., 1974*a*). These changes are consistent with the hypothesis that anions affect the ratio of membrane-bound Ca to ionized Ca (Carvalho, 1968). Specifically, if an anion substitution, such as Cl for propionate, reduces the large fraction of Ca that is apparently bound to the inner surface of the SR (see Vanderkooi and Martonosi, 1971), then tension can be expected. Following this reduction of bound Ca, the concentration of ionized Ca within the SR should increase and Ca efflux should result because of the increase in the ionized Ca gradient from the inside to the outside of the SR, and because an increase in ionized Ca within the SR inhibits Ca uptake (see Weber, 1968).

The Effects of Mg and Procaine

Transient tensions can also be induced without changing anions. Before eliciting the second tension in Fig. 6 (*top center*), a skinned fiber was exposed to an excess concentration of free Mg during the initial Ca-loading procedure. A transient tension was induced by removing the excess Mg from the bath. These Mg "withdrawal" tensions could be induced in either Cl or propionate bathing solutions and therefore were not dependent upon the anion. Before discussing the possible mechanisms by which Mg might affect Ca distribution, other experimental observations will be described.

Excess free Mg attenuates or blocks the anion-induced transient tensions in human (Wood, Reuben, and Brandt, 1974), crayfish (*unpublished observations*), and frog (Stephenson and Podolsky, 1973) skinned fibers. Mg also attenuates and in some cases blocks caffeine-induced tensions in both crayfish (*unpublished observations*) and human skinned fibers (Wood et al.,

1974). The effects of increasing the ionized Mg concentration on the Ca-regulating systems are most striking after the fiber is in tension (Fig. 7, *bottom tracings*). Once the fiber is in tension, addition of Mg causes transient relaxation. This contrasts with the responses of resting and Brij-treated fibers, in which elevation of free Mg induces a long-lasting increase in tension; the transient tension evoked by reducing Mg is absent after Brij treatment and is replaced by a decrease in tension (Fig. 6, *top tracings* and Fig. 7, *bottom tracings*).

The modification of tension by Mg in skinned fibers with intact SR is mimicked by procaine (5 to 10 mM) (Wood et al., 1974). The similar effects of Mg and procaine on tension and on the ATPase activity of isolated SR suggest to us that both act by reducing Ca efflux from the SR without attenuating ability of the SR to take up Ca. The relevant observations on isolated SR are: (a) free Mg, or procaine, attenuates caffeine-induced loss of Ca from isolated SR vesicles (Weber and Herz, 1968); (b) procaine (3.3 mM) does not stimulate the ATP-dependent uptake of Ca by SR (Weber, 1968; Weber and Herz, 1968); (c) free ATP, but not free Mg, stimulates uptake of Ca by isolated SR (Weber, 1968).

Mg and procaine may enhance Ca accumulation by reducing the Ca

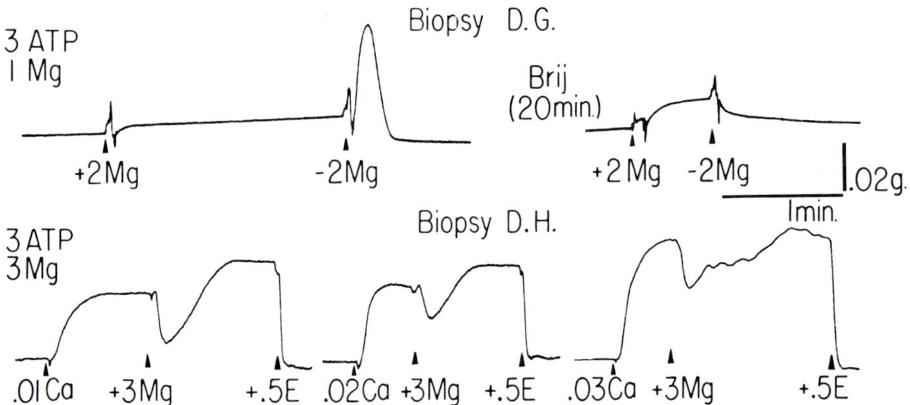

FIG. 7. The effects of Mg on both relaxed and contracted human fibers. *Top tracings:* The skinned fiber was preexposed to Ca in a solution containing 3 mM ATP and 1 mM Mg. A small slow-rising tension was induced when the Mg concentration was increased to 3 mM. Reduction of the Mg concentration to the 1-mM level evoked a nearly maximum transient tension. The fiber was then treated with Brij for 20 min before repeating the step change in the Mg concentration. The response to adding 2 mM Mg was larger after destroying the Ca-accumulating properties of the SR, and the transient tension following the reduction of Mg was abolished. *Bottom tracings:* In a second fiber from another biopsy, tensions were elicited by the addition of 0.01, 0.02, and 0.03 mM Ca to the bathing solution. Tensions were terminated by adding 0.5 mM EGTA (E) to the bath. The Mg concentration was increased from 3 mM to 6 mM in each case after an apparent steady tension was attained. A transient relaxation followed the addition of Mg; the lower the Ca concentration, the larger the relaxation.

permeability of a membrane-bound compartment such as the SR. This hypothesis is consistent with the observations on isolated SR. Accordingly, skinned-fiber tensions induced by reducing Mg or procaine concentrations in the bath are probably due to increased permeability of these membranes to Ca. In turn, a stimulus such as Cl or caffeine that normally causes a net efflux of stored Ca may be partly or totally counteracted by the opposing effect of decreased Ca permeability caused by excess Mg or procaine. Finally, the observation that Mg or procaine makes fibers in tension relax is predicted by, and consistent with, the hypothesis that those agents attenuate Ca efflux and thereby increase the net uptake of Ca by SR.

The addition or removal of Mg from the solution bathing a detergent-treated fiber may evoke a step change in tension. This change in tension occurs only after preexposure to Ca and is suppressed by adding a Ca chelator such as EGTA. These data suggest that in skinned fibers without a functional SR the effects of Mg are caused by displacement of Ca from sites (X) other than troponin sites. The displaced Ca then becomes available to troponin and, if the quantity of displaced Ca is sufficient, an increase in tension occurs (Fig. 6). This long-lasting increase in tension may not be observed in skinned fibers with a functional SR because of the rapid removal of matrix Ca by an unloaded or partially loaded SR.

Effects of Caffeine

The third tension of Fig. 6 was evoked by adding caffeine (10 mM). This tension was larger than the preceding ones and attained p_o even though the period allowed for loading the Ca-sequestering systems was identical. Based on studies of isolated SR, it has been proposed that caffeine causes a net loss of Ca from the SR by blocking Ca uptake (Weber, 1968). Bianchi (1961), however, studying intact muscle, concluded that caffeine increased Ca permeability and caused net efflux of Ca from the SR.

The data of Fig. 8, analyzed in conjunction with the actions of Mg and procaine, do not support the conclusion that the major effect of caffeine is to attenuate Ca uptake. The two tensions in the upper portion of Fig. 8 were elicited by increasing the Ca concentration to 20 μM and 10 μM in a solution containing 40 mM caffeine. The two tensions at the bottom were also evoked in 40 mM caffeine and by the same two levels of Ca, but after treating the fiber with Brij-58. Similar tension records were obtained in chemically skinned human fibers. After attaining a steady tension (less than p_o), Mg and procaine were added sequentially. In the first recording 10 mM procaine induced a small transient relaxation, followed by an even larger transient when 5 mM free Mg was added with the procaine. The order of adding Mg and procaine was then reversed in the second tension tracing. The final two records show that the transient relaxations were abolished after destroying the Ca-accumulating ability of the SR by Brij-58. With 40 mM caffeine in the

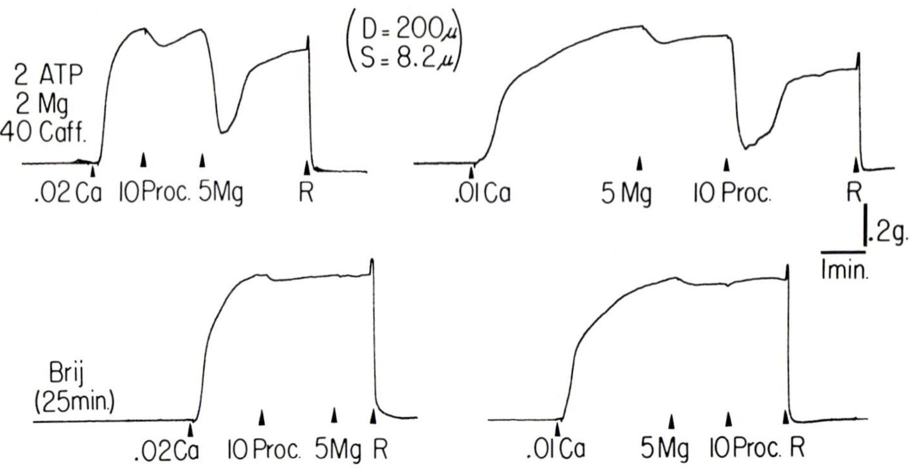

FIG. 8. The effects of procaine, Mg, and caffeine on a Ca-induced tension. The four tensions were evoked by addition of either 0.01 or 0.02 mM Ca to a bathing solution containing 40 mM caffeine. *Top left:* Transient relaxations occurred upon adding first procaine and then Mg to the solution bathing a maximally activated fiber. Solution R terminated the tension. *Top right:* Similar effects on a tension induced by lower Ca and by reversal of the order of application of Mg and procaine. *Bottom tracings:* The same procedure was repeated after treating the fiber for 25 min with Brij-58 (0.5%). The transient relaxations were absent when Mg and procaine were applied separately or together. Note that the tensions induced by 0.01 mM Ca were slower to rise than those induced by 0.02 mM Ca. Further description in text.

bathing solutions, the fiber was able to accumulate Ca when procaine and Mg were added. Although final proof must await further studies, the competitive relation between caffeine and Mg (or procaine) supports the suggestion that caffeine increases the Ca permeability of the skinned fiber's internal membranes without blocking Ca uptake.

Effects of Ca

One of the most intriguing hypotheses about Ca regulation in skinned fibers is that Ca promotes the release of Ca from the SR (Endo et al., 1970; Ford and Podolsky, 1970, 1972). Experiments were designed to test that hypothesis (Fig. 9). Skinned crayfish fibers were exposed to a subthreshold level of buffered Ca (pCa 6.7; EGTA total = 5 mM) for a fixed time (5 min). The Ca-EGTA solution was followed with a wash (W) solution containing a series of increasing levels of EGTA and no added Ca. When the Ca-EGTA solution was washed out with one containing 0.3 mM free EGTA, or less, a large tension resulted. That was terminated after 1 min by washing with solution R containing 5 mM EGTA. After a delay tensions developed rapidly. Removal of the Ca buffer, followed by addition of free EGTA, must cause a drop in the concentration of ionized Ca within the fiber. During a

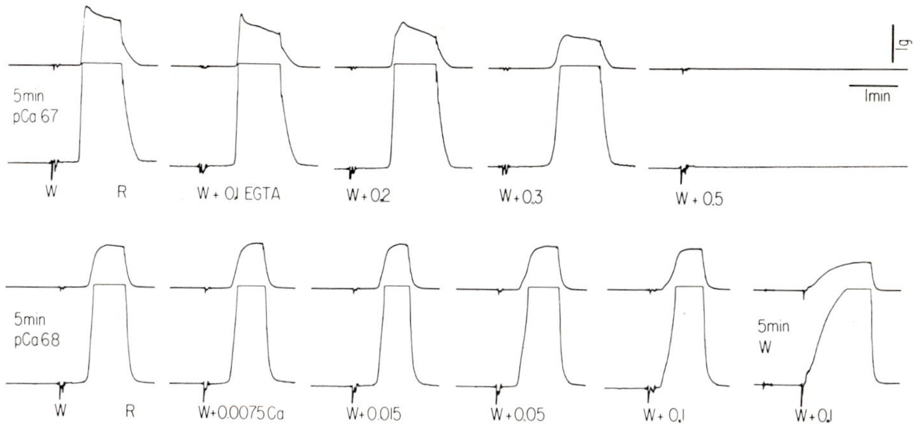

FIG. 9. Stored-Ca mobilization initiated by a change in the Ca-buffering capacity of the bathing medium. *Top set of tracings:* Low- and high-gain records of four sequential tensions. Before each tension the fiber was preexposed for 5 min to a CaEGTA/EGTA buffer solution (pCa 6.7). The Mg and ATP concentrations were kept at 2 mM throughout the experiments. The tensions developed after a delay period following the removal of the buffer from the bathing medium (*artifacts mark the solution change*). The 2nd, 3rd, and 4th tensions progressively decreased in amplitude as the concentrations of free EGTA in the wash solution increased. No tension occurred in the presence of 0.5 mM EGTA. *Bottom set of tracings:* Fiber was preexposed for 5 min to a buffered pCa 6.8 solution before each tension, with the exception of the last one, and the wash solutions contained increasing amounts of Ca up to 0.1 mM. The last tension was evoked by exposing the fiber to 0.1 mM Ca without a preexposure to buffered Ca. Further description in text.

period of falling levels of ionized Ca, tension is elicited. Although we cannot measure directly the quantity of Ca released that causes a tension, the induced tension was not blocked until 0.5 mM EGTA was present in the wash solution.

When the ionized Ca concentration in the bath was increased (*lower recordings,* Fig. 9), it did not affect the development of tension following removal of Ca-EGTA. The fiber was again exposed to a buffered Ca solution for 5 min (pCa 6.8; EGTA total = 5 mM). After the first response, variable amounts of free Ca, rather than free EGTA, were added to the solutions used to wash out the Ca buffer. Addition of up to 0.015 mM Ca (*fourth tension*) or 0.1 mM Ca (*fifth tension*) reduced the delay and added a foot to the initial phase of tension development. These two effects can be explained from the response of the fiber to 0.1 mM Ca alone, without a 5-min preexposure to the buffered Ca solution (*sixth tension*). Without the Ca preexposure, the SR can be loaded only minimally and is therefore capable of rapidly sequestering Ca. Consequently, upon addition of 0.1 mM Ca solution, Ca diffusing into the fiber must be divided among SR, troponin, and non-troponin Ca-binding sites. The slow rise of the sixth tension and the foot of the fifth tension probably result from direct activation of the outermost annulus of myofilaments by Ca diffusing inward.

The fast-rising component in all of the tensions induced after preexposure to Ca was not affected by increasing ionized Ca up to 0.1 mM in the bath. Obviously the fast component of tension must involve mobilization of Ca stored within the fiber during the preexposure to Ca. Since this fast component was not affected by adding either free Ca (up to 0.1 mM) or free EGTA (up to 0.3 mM) to the bath, Ca does not seem to initiate release of Ca from the SR.

SUMMARY

Our studies are summarized in Fig. 10. Drawing upon a pool of ionized Ca are the SR, troponin sites, and non-troponin binding sites (X). The ionized Mg pool is larger than that for Ca, in keeping with the greater concentrations of Mg in the experimental solutions. The Mg pool provides Mg to the X sites, where Mg binds competitively with Ca. The dotted arrow running from the Mg pool to the Ca efflux arrow reflects our view that Mg, like procaine, affects Ca efflux from the SR. While Mg modifies Ca distribution within the skinned fiber by displacing Ca from X and by reducing Ca efflux from the SR, caffeine and procaine act primarily by altering the Ca perme-

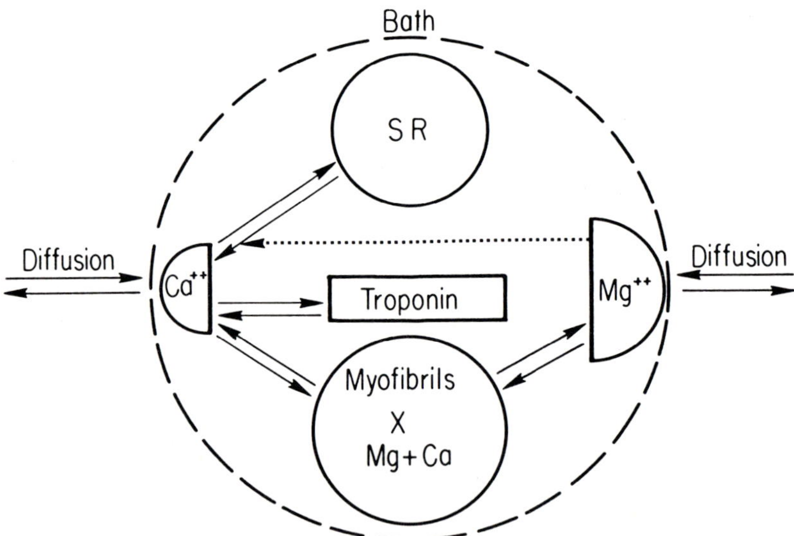

FIG. 10. Diagram of Ca and Mg distribution within a skinned fiber. The boundary between the myofilaments and the bathing solution is indicated by the circular broken line. The areas enclosed by semicircles represent the ionized pools of Ca and Mg that diffuse into the fiber after removal or disruption of the surface membrane. These pools provide Ca and Mg to the SR, X sites, and troponin. The SR does not sequester Mg and troponin's affinity for Mg is relatively low. Dotted arrow indicates that Mg hinders Ca efflux from the SR.

ability of the SR membranes. Anion-induced tensions can be explained by experiments which show that anions modify Ca binding by the SR as well as other membranes.

In relating the observations on Ca regulation in skinned fibers to what may occur in intact fibers, the implications of the anion data and the demonstrated existence of the X sites are important. If, for example, the excitatory signal leads to a shift of Ca from a membrane-bound state to an ionized state within the SR lumen, then net efflux of ionized Ca into the filament space would follow, since a rise in ionized Ca within the lumen increases the gradient for Ca efflux and inhibits the ATPase activity of the SR's Ca uptake system (Weber, 1968). Ca leaving the SR would bind to both troponin and X sites. Once the SR releases its Ca load, the ATPase activity of the SR is no longer inhibited. Ca bound to troponin sites and X sites is then relinquished to the rapid sequestering system of the SR. The Ca buffering provided by the X sites should prolong the time in which Ca is available to troponin and reduce Ca loss via the surface membrane (see Reuben et al., 1974*a*) by keeping the ionized Ca at a low level. Furthermore, Ca buffering by the X sites explains the observation that during peak tension the free-Ca level within the matrix of intact fibers is lower than expected (Ashley and Ridgway, 1970; Rudel and Taylor, 1973).

ACKNOWLEDGMENTS

This work was supported in part by funds from the Muscular Dystrophy Associations of America, Inc.; by Public Health Service research grants NS 03728, NS 05910, GM 18640, and HL 16082; training grant NS 05328 and grant NS 11766 from the National Institute of Neurological Diseases and Stroke; and grant GB 31807X from the National Science Foundation. D. S. W. and J. R. Z. are trainee fellows on grant NS 05328.

REFERENCES

Ashley, C. C., and Ridgway, E. B. (1970): On the relationships between membrane potential, calcium transient and tension in single barnacle muscle fibres. *J. Physiol.* 209:105–130.

Bianchi, C. P. (1961): The effect of caffeine on radiocalcium movement in frog sartorius. *J. Gen. Physiol.* 44:845–858.

Brandt, P. W., Reuben, J. P., Girardier, L., and Grundfest, H. (1965): Correlated morphological and physiological studies on isolated single muscle fibers. I. Fine structure of the crayfish muscle fiber. *J. Cell Biol.* 25:233–261.

Brandt, P. W., Reuben, J. P., and Grundfest, H. (1972): Regulation of tension in the skinned crayfish muscle fiber. II. Role of calcium. *J. Gen. Physiol.* 59:305–317.

Carvalho, A. (1968): Calcium-binding properties of sarcoplasmic reticulum as influenced by ATP, caffeine, quinine and local anesthetics. *J. Gen. Physiol.* 52:622–642.

Costantin, L. L., Franzini-Armstrong, C., and Podolsky, R. J. (1965): Localization of calcium-accumulating structures in striated muscle fibers. *Science* 147:158–159.

Costantin, L. L., and Podolsky, R. J. (1967): Depolarization of the internal membrane system in the activation of frog skeletal muscle. *J. Gen. Physiol.* 50:1101–1124.

Davis, W. L., Matthews, J. L., and Martin, J. H. (1974): An electron microscopic study of myo-filament calcium binding sites in native, EGTA-chelated and calcium reloaded glycerolated mammalian muscle. *Calcif. Tiss. Res.* 14:139–152.

de Meis, L., Hasselbach, W., and Machado, R. D. (1974): Characterization of calcium oxalate and calcium phosphate deposits in sarcoplasmic reticulum. *J. Cell Biol.* 62:505–509.

Ebashi, S., and Endo, M. (1968): Calcium ion and muscle contraction. *Prog. Biophys. Mol. Biol.* 18:123–183.

Endo, M., Tanaka, M., and Ogawa, Y. (1970): Calcium induced release of calcium from the sarcoplasmic reticulum of skinned skeletal muscle fibres. *Nature* 228:34–36.

Fiehn, W., and Hasselbach, W. (1970): The effect of phospholipase A on the calcium transport and the role of unsaturated fatty acids in ATPase activity of sarcoplasmic vesicles. *Eur. J. Biochem.* 13:510.

Ford, L. E., and Podolsky, R. J. (1970): Regenerative calcium release within muscle cells. *Science* 167:58–59.

Ford, L. E., and Podolsky, R. J. (1972): Intracellular calcium movements in skinned muscle fibres. *J. Physiol.* 223:21–23.

Hasselbach, W. (1964): Relaxing factor and the relaxation of muscle. *Prog. Biophys. Mol. Biol.* 14:169–222.

Hasselbach, W., and Makinose, M. (1961): Die Calciumpumpe der "Erschlaffungsgrana" des Muskels und ihre Abhängigkeit von der ATP-Spaltung (summary in English). *Biochem. Z.* 333:518–528.

Hasselbach, W., and Makinose, M. (1963): Uber den Mechanismus des Calciumtransportes durch die Membranen des sarkoplasmatischen Reticulums (summary in English). *Biochem. Z.* 339:94–111.

Hasselbach, W., Makinose, M., and Fiehn, W. (1974): Activation and inhibition of the sarco-plasmic calcium transport. In: *Calcium and Cellular Function,* edited by A. W. Cuthbert, pp. 75–84. St. Martin's Press, New York.

Hellam, D. C., and Podolsky, R. J. (1969): Force measurements in skinned muscle fibres. *J. Physiol.* 200:807–819.

Hober, R., editor (1945): *Physical Chemistry of Cells and Tissues.* Blakiston, Philadelphia.

Hodgkin, A. L., and Horowicz, P. (1960): The effect of nitrate and other anions on the mechani-cal response of single muscle fibres. *J. Physiol.* 153:404–412.

Natori, R. (1954): The property and contraction process of isolated myofibrils. *Jikei Kai Med. J.* 1:119.

Orentlicher, M., Reuben, J. P., Grundfest, H., and Brandt, P. W. (1974): Calcium binding and tension development in detergent-treated muscle fibers. *J. Gen. Physiol.* 63:168–186.

Reuben, J. P., Brandt, P. W., Berman, M., and Grundfest, H. (1971): Regulation of tension in the skinned crayfish muscle fiber. I. Contraction and relaxation in the absence of Ca (pCa > 9). *J. Gen. Physiol.* 57:385–407.

Reuben, J. P., Brandt, P. W., and Grundfest, H. (1967): Tension evoked in skinned crayfish muscle fibers by anions, pH, and drugs. *J. Gen. Physiol.* 50:2501.

Reuben, J. P., Brandt, P. W., Katz, G. M., and Grundfest, H. (1970): Augmentation of re-sponses to calcium injection by agents that reduce calcium sequestration. *J. Gen. Physiol.* 55:140.

Reuben, J. P., Brandt, P. W. and Grundfest, H. (1974a): Regulation of myoplasmic calcium concentration in intact crayfish muscle fibers. *J. Mechanochem. Cell Motility,* 2:269–285.

Reuben, J. P., Brandt, P. W., Sorenson, M. M., and Eastwood, A. (1974b): Regulation of con-tractile activity in isolated "skinned" human muscle fibers. *Fed. Proc.* 33:1260, Abst. 202.

Rudel, R., and Taylor, S. R. (1973): Aequorin luminescence during contraction of amphibian skeletal muscle. *J. Physiol.* 233:5P.

Sollner, K. (1949): The origin of bi-ionic potentials across porous membrane of high ionic selectivity. *J. Phys. Chem.* 53:1211–1226.

Stephenson, E. W., and Podolsky, R. J. (1973): Regulation by Mg of Ca movement in skinned muscle fibers. *Fed. Proc.* 32:373.

Vanderkooi, J. M., and Martonosi, A. (1971): Sarcoplasmic reticulum. XIII. Changes in the fluorescence of 8-Anilino-I-Naphalene sulfonate during Ca^{2+} transport. *Arch. Biochem. Biophys.* 144:99–106.

Weber, A. (1968): The mechanism of action of caffeine on sarcoplasmic reticulum. *J. Gen. Physiol.* 52:760–772.

Weber, A., and Herz, R. (1968): The relationship between caffeine contracture of intact muscle and the effect of caffeine on reticulum. *J. Gen. Physiol.* 52:750–759.

Winegrad, S. (1971): Studies of cardiac muscle with a high permeability to calcium produced by treatment with ethylenediaminetetraacetic acid. *J. Gen. Physiol.* 58:71–93.

Wood, D. S., Reuben, J. P. and Brandt, P. W. (1974): The effects of Mg and procaine on Ca redistribution within chemically skinned human muscle fibers. *The Physiologist* 17:361.

Concepts of Membranes in Regulation and Excitation,
edited by M. Rocha e Silva and G. Suarez-Kurtz.
Raven Press, New York © 1975.

Is a Ca Current Necessary for Excitation-Contraction Coupling in Skeletal Muscle?

Guilherme Suarez-Kurtz and John P. Reuben

Department of Pharmacology, Institute of Biomedical Sciences, Federal University of Rio de Janeiro, Brazil and Laboratory of Neurophysiology, Department of Neurology, College of Physicians and Surgeons, Columbia University, New York 10032

INTRODUCTION

In a number of biological systems extracellular Ca is necessary for, and in some cases has been shown to be directly involved in, the conveyance of information across cellular membrane (see Rasmussen, 1970). A correlation between the magnitude of an inward Ca- current across the presynaptic membrane and the quantity of transmitter released has been established (Kusano, 1968; Katz and Miledi, 1969) and it serves as an example of the direct involvement of Ca in a signaling process.

Although most muscle physiologists agree that extracellular Ca is necessary for the signaling of the contractile elements in muscle, the mode of its involvement in the excitation process has been a matter of considerable dispute. On the one hand, extracellular Ca is believed to be necessary only insofar as it maintains the electrical properties of the muscle membrane (see Ishiko and Sato, 1957; Jenden and Reger, 1963; Edman and Grieve, 1964; Sandow, 1965). On the other hand, extracellular Ca is thought to be required by the muscle fiber — as it is at presynaptic terminals — for transporting inward current that serves as an excitatory signal. Proponents of the latter view must be concerned with the determination of the quantity of Ca entering the muscle fiber during excitation, since Ca supplied in sufficient quantities to the myoplasm will activate the contractile system (see Heilbrun and Wiercinski, 1947; Portzehl, Caldwell, and Ruegg, 1964). By contrast, at the presynaptic terminal, where a membrane Ca current is necessary for transmitter release, the injection of Ca into the cytoplasm does not cause a release of transmitter (Miledi and Slater, 1966; Kusano, 1968).[1] The quantities of Ca entering peripheral striated muscles during excitation have been measured (Bianchi, 1968) and they are too small to completely activate the

---

[1] In a recent abstract [*Proc. Roy. Soc. London (B)*, 1973, 183:421–425], Miledi noted the release of small quantities of transmitter during injection of Ca.

contractile system (Bianchi, 1969). Although these measurements do not preclude the possibility of a direct involvement of a membrane Ca current in the excitation process, they served to discourage, for some time, further consideration of the hypothesis.

As an outgrowth of studies on the properties of the transverse tubular system (TTS) of crayfish muscle fibers a transmembrane current was again suggested as a necessary step in the signaling process (Girardier, Reuben, Brandt, and Grundfest, 1963). Specifically, the current channeled through the TTS and flowing across the diadic membranes was thought to be part of the excitation process. Additional studies on the excitation process in crustacean muscle fibers (Reuben, Brandt, Garcia, and Grundfest, 1967; Suarez-Kurtz, Reuben, Brandt, and Grundfest, 1972) led the authors to conclude that the diadic current transported by Ca is the critical one for excitation. This rebirth of a hypothesis in a somewhat modified form has not been based solely on data obtained from striated crustacean fibers. Recent work by Chiarandini and Stefani (1973) ascribes the action of Mn in attentuating tensions in frog twitch fibers to blocking of the membrane Ca current. In part, the renewal of interest in a transmembrane Ca current has come from the suggestion made by several investigators that Ca entering the muscle fiber releases additional Ca stored within the SR (Bianchi and Bolton, 1967; Reuben, Katz, and Berman, 1969; Endo, Tanaka, and Ogawa, 1970; Ford and Podolsky, 1972). Such a mechanism—coupled with the localization of the inward Ca current at the diadic or triadic level—offers an explanation for the basic steps of the excitation process. Whether or not the postulated "Ca-releases-Ca" mechanism is operative under physiological conditions (see Reuben, Brandt, and Grundfest, 1974) should not distract further exploration of excitatory signaling via an inward Ca current in muscle.

Here we present some findings on the dependence of tension generation in so-called fast muscle fibers from the crab *Callinectes danae* on membrane Ca conductance. A comparison is drawn between the new data and data obtained from earlier experiments on "slow" crayfish fibers. By convention, a fiber is called "fast" if it generates a conducted action potential and a twitch while bathed in control saline. We say "fast" because of the speed of contraction and also because twitch kinetics in crab fibers, although induced by treatment, are similar to those in frog fast fibers.

CORRELATION BETWEEN THRESHOLDS FOR ELICITING TENSION AND CA SPIKES

Crayfish muscle fibers dissected from the muscles within the meropodite of the walking limb respond to depolarization with graded membrane electrogenesis and graded contractions. All-or-none Ca spikes accompanied by contractions can, however, be elicited in these fibers when they are exposed to procaine (Takeda, 1967), tetraethylammonium (Fatt and Ginsborg, 1958), caffeine (Chiarandini, Reuben, Brandt, and Grundfest, 1970), or several divalent cations (Fatt and Ginsborg, 1958). The induction of Ca-spike electrogenesis by these agents probably results from a relative increase in membrane Ca conductance, caused by enhanced Ca conductance or decreased K conductance during depolarization. In a recent publication (Suarez–Kurtz et al., 1972) we presented data suggesting that an increase in membrane Ca conductance, leading to an inward Ca current, is an essential step in excitation-contraction coupling (ECC).

The experiments, performed on isolated fibers of the crayfish *Orconectes virilis,* were designed so that the threshold membrane depolarization required for initiating a just-detectable tension could be compared under similar conditions to the threshold for inducing Ca spikes. Procaine was used to convert the graded response to all-or-none Ca spikes because the effects of procaine can be completely and rapidly reversed. After measuring the level of depolarization required to initiate tension, procaine (1 mg/ml) was added to the bathing medium and the threshold for eliciting a Ca spike was recorded. The tension threshold was unchanged (± 1.5 mV) after washing out the procaine. Tension and spike thresholds were measured in fibers exposed to a series of concentrations of the divalent cations Ca, Mg, Mn, and Ni, and of the lyotrophic anion SCN. The experimental solutions were isotonic, the divalent cations were substituted for Na, and SCN replaced Cl.

Figure 1 is composed of microelectrode recordings of membrane potentials and membrane current and strain gauge measurements of tension and the derivative of tension. The distance between the points of impalement of the fiber by the voltage and current microelectrodes was kept at a value well below the diameter of the fiber. This practice minimized possible effects of length constant (λ) changes on the measurement of the voltage displacements required for the initiation of tension and spikes. However, this consideration was seldom important, for under most experimental conditions, such as that used in collecting the data of Fig. 1 (3 mM Mn), the effective resistance (R_E) and λ did not change. While the latter parameters were not affected in fibers exposed to 3 mM Mn, the threshold for both tension and the Ca spike shifted to lower values of membrane potentials. For comparison, changes in the threshold values are given in terms of the incremental depolarization from the resting potential, rather than the absolute potential. Measurements were made of the incremental depolarization required for initiation of tension (Δt) and the Ca spike (Δs) in fibers exposed to a series of concentrations of each of the previously mentioned ions. A plot of Δt versus Δs was constructed from data gathered from experiments on 15 single fibers (Fig. 2). The regression line, calculated by the method of least squares was: $\Delta t = 1.23\Delta s + 2.96$. The correlation coefficient (r) was 0.95 and the level of significance of the correlation (p) was

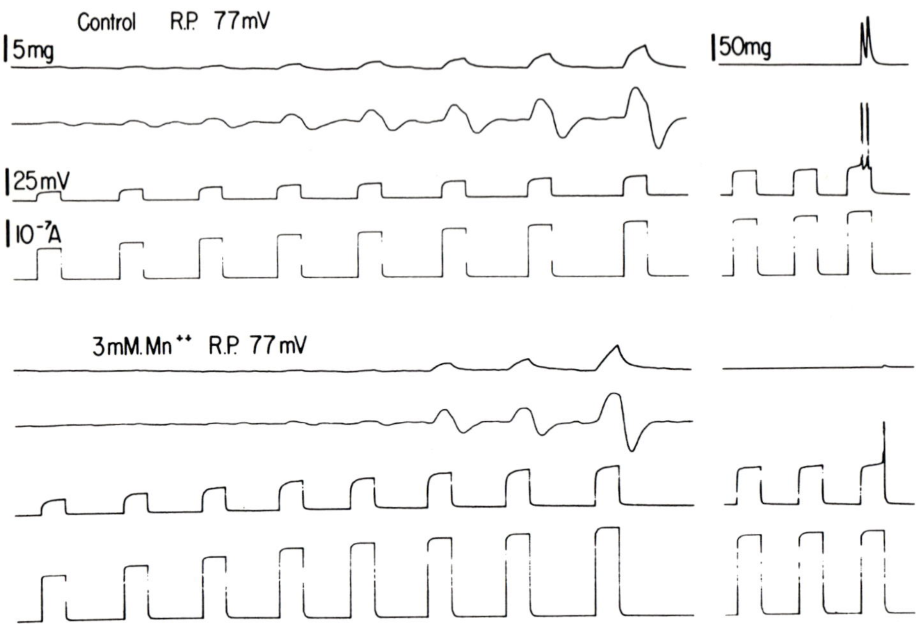

FIG. 1. Determination of tension and spike threshold and their modifications by 3 mM Mn. *Top left 4 rows:* Measurements of tension, the derivative of tension, membrane potential, and transmembrane current obtained from a crayfish fiber bathed in control saline. The depolarizing current pulses of increasing amplitude each last 1 sec. *Top right 3 rows:* The corresponding recordings (without the derivative of tension) after treating the fiber with procaine (1.0 mg/ml). *Bottom left:* After exposing the fiber to 3 mM Mn, a larger depolarization is required to elicit a just-detectable tension. *Bottom right:* The fiber was again exposed to procaine, but in the presence of 3 mM Mn. The threshold for evoking Ca spikes was also elevated. Note the large attenuation of the tension during the spike in the presence of Mn. R.P. is resting potential. (From Suarez-Kurtz et al., 1972.)

greater than 0.95. This significant correlation between thresholds for inducing tension and for initiation of an inward Ca current (Ca spike) provides the basis for the conclusion that the inward Ca current is the excitatory signal (Suarez-Kurtz et al., 1972).

It would have been preferable, if technically possible, to measure the Ca current directly and compare it with the evoked tension. Our attempt to measure tension and inward Ca current simultaneously by voltage clamp techniques was not successful. The limiting factor was the large increase in membrane conductance as the membrane potential was reduced toward zero. Under this condition the length constant was reduced from about 1.5 mm to as little as 0.5 μ (Brandt, Orentlicher, Katz, *unpublished observations*). In the face of the variation of λ, the membrane potential could not be held constant over the length of the contracting fiber. That is unfortunate, for an immediate and direct test of the hypothesis would be to uniformly depolarize a fiber to or above the Ca equilibrium potential (E_{Ca})

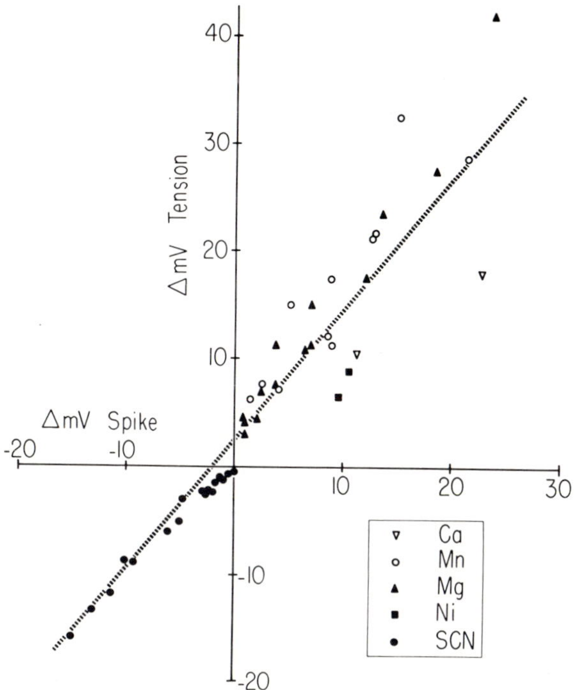

FIG. 2. Correlation between the depolarization required to elicit Ca spikes (*abscissa*) and tension (*ordinate*). Symbols are identified in inset. Further description in text. (From Suarez-Kurtz, et al., 1972.)

and to determine whether or not tension is elicited $\left(E_{Ca} = \dfrac{RT}{zF} \ln \dfrac{Ca_{out}}{Ca_{in}}\right)$. If tension is, indeed, initiated by an inward Ca current, then tension should not develop when the membrane potential $\geq E_{Ca}$. This, of course, assumes that the inward Ca current is determined by the relationship $I_{Ca} = g_{Ca}$ $(E_M - E_{Ca})$, where I_{Ca} is the membrane Ca current, g_{Ca} is the conductance of the membrane towards Ca, and E_M is the membrane potential.

Further evidence, albeit indirect, supporting the view that an inward Ca current is essential for ECC comes from experiments using the so-called fast muscle fibers of the South American crab *Callinectes danae*.

DEPENDENCE OF FAST-CONTRACTION KINETICS ON A RAPID MEMBRANE CA ACTIVATION IN CRAB FIBERS

The procedures for isolating the large crab fibers (150 to 160 μ in diameter), mounting them in experimental chambers, and recording tension and membrane parameters were previously described (Suarez-Kurtz et al.,

1972). The concentrations of the salts in the crab saline (NaCl = 460 mM; KCl = 5 to 12.5 mM; CaCl$_2$ = 12.5 mM; MgCl$_2$ = 9.5 mM) were higher than those in crayfish saline. In most crab fiber experiments the lower K concentration was used (5 mM), as in crayfish saline. The resting potential of fibers bathed in 5 mM K saline was about 20 mV higher than fibers bathed in 12.5 mM K solutions, whose membrane potential was about −60 mV.

The characteristic membrane potential changes elicited by a 5 msec depolarizing current pulse are shown in Fig. 3A. The fiber was bathed in the 5 mM K saline, but similar results were obtained when fibers were bathed in 12.5 mM K solutions (see Fig. 5, *top left tracings*). An exponential time course was observed for depolarization less than 40 mV; for larger depolarization, a brief (about 2 msec) initial transient was noted (Fig. 3A, B, and D). Under a high-power microscope, the fiber contracted rapidly in the region surrounding the current microelectrode when, and only when, the initial transient appeared in the voltage traces. The local tensions, although small (usually less than 3% p_o), could be recorded (see Figs. 4 and 5). The rapid contraction of these fibers was best recorded after they had been treated to respond to depolarization with all-or-none spikes. In treated fibers, tension may reach 50% p_o in 20 to 50 msec.

We have concluded from the results of experiments such as those of Fig. 3 that the rapid potential transients are due to a rapid change in the Ca conductance of the membrane. The fact that the transients are not modified in fibers exposed to 10^{-6} g/ml tetrodotoxin (TTX; Fig. 3B) or blocked by substituting tris(hydroxymethyl)aminomethane for the Na of the

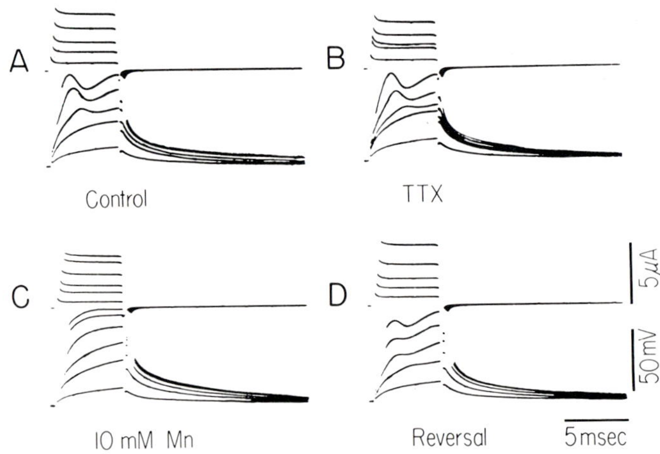

FIG. 3. Effect of Mn and TTX on the membrane responses to depolarizing pulses in a single crab muscle. The baseline of the current recording indicates zero membrane potential. *A*, control responses; *B* and *C*, recordings taken 5 min after exposure to 10^{-6} g/ml TTX and 10 mM Mn, respectively; *D*, recording taken 5 min after removal of Mn. Further description in text.

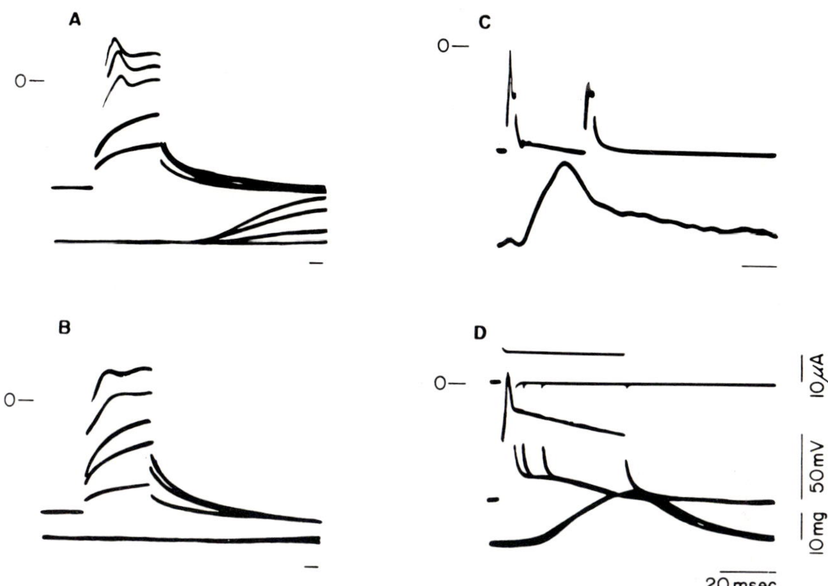

FIG. 4. Potential and tension recordings taken from two different crab muscle fibers (*A, B* and *C, D*) bathed in saline containing 5 mM K. *A:* Membrane and contractile response elicited by depolarizing pulses (25 msec duration) in a fiber bathed in the control saline; *B:* Membrane and response after the addition of 10 mM Mn. *C:* Inhibition of the rapid membrane activation and tension development with repetitive stimulation. *D:* Effect of varying the duration of the depolarizing current pulses (6, 10, 20, and 50 msec) on the membrane and contractile responses. Note that the amplitude and time course of the contraction were not affected by increasing the pulse duration beyond that necessary to elicit the rapid membrane conductance change.

saline rules out a dependence of their generation on the membrane conductance toward Na. The abolition of the potential transients—and, as will be described shortly, the abolition of tension—by 10 mM Mn (Fig. 3C) is presumptive evidence that the transients arise because of a change in the Ca conductance of the membrane. Mn blocks the increase in membrane Ca conductance during depolarization in a variety of membranes (see Suarez-Kurtz et al., 1972). The transients reappeared after a brief washing of the fiber with control saline (Fig. 3D).

In Fig. 4 both the membrane and tension responses of another fiber are recorded before (A) and after (B) exposing the fiber to 10 mM Mn. The potential transients did not appear until the membrane potential decreased by more than 65 mV. In spite of this large depolarization, tension was not observed or recorded. Further depolarization of the membrane elicited both transients and tensions. The rising phase of the three tensions (lower trace of Fig. 4A) were elicited by and corresponded with the three potential transients. After treatment of the fiber with 10 mM Mn, the transient po-

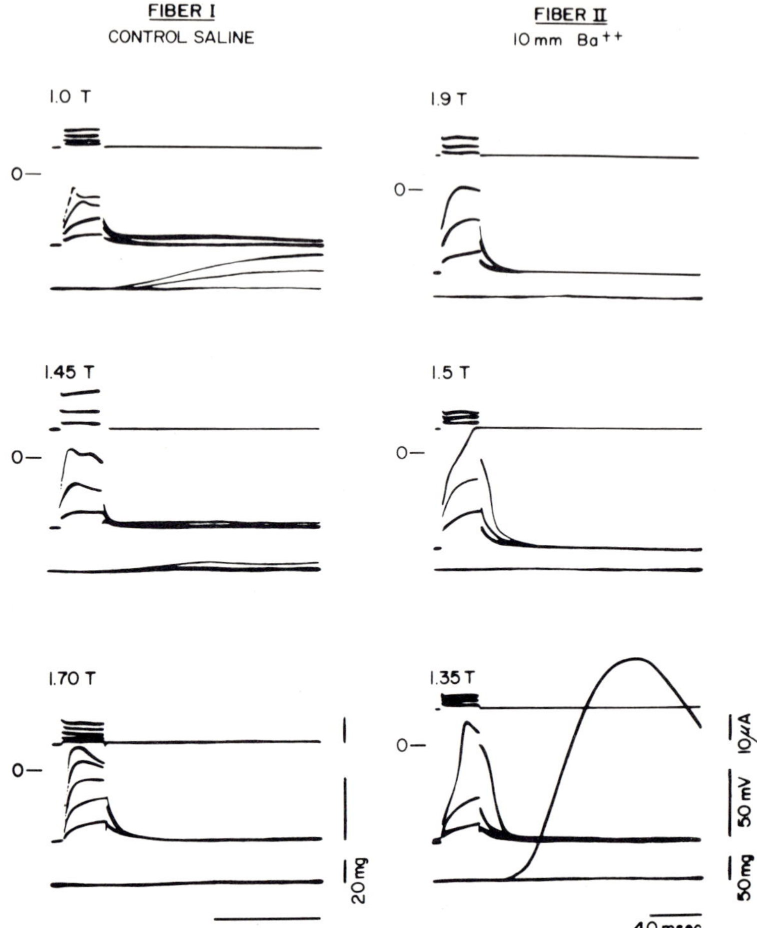

FIG. 5. Inhibitory effects of hypertonic solutions. *Fiber I:* Inhibition of the membrane Ca activation and tension by increasing the tonicity of the bathing medium with Tris propionate. *Fiber II:* Blockade of Ba-induced spikes and twitches by hypertonic solutions. All records were taken 5 min after the fibers were soaked in the indicated solutions.

tential changes were depressed and tension was not recorded (Fig. 4B). In some fibers, attenuated transients persisted in the presence of 10 mM, and a visible, but not recordable, contraction also persisted. Both the residual membrane and contractile responses were abolished by increasing the Mn concentration to either 15 or 20 mM.

Other characteristics of the responses of the fast crab fibers to depolarization are shown in Fig. 4C and D. There was a refractory period; a second potential transient cannot be evoked for 30 to 50 msec following an initial transient response (Fig. 4C). During this refractory period, in which brief

Ca activation could not be elicited, complete depolarization of the membrane did not induce tension. This lack of dependence of tension and its initiation on the level of membrane potential per se was also apparent in another experiment (Fig. 4D). The duration of the depolarizing pulse was increased in four steps from 6 to 50 msec. The four corresponding tensions (*lower tracings,* Fig. 4D) appeared as one superimposed record of the same tension. The reproducibility of these tensions, despite the increasing period of depolarization, was attributed to the constancy of the initial transient.

The slow decline of potential during the 50-msec pulse was probably due to delayed increase in conductance following the initial transient. A delayed increase in membrane conductance could also account for the depolarizing after-potentials that persisted after cessation of the current pulse for the three short-duration pulses. The conductance increase, however, would have to involve an ion or ions with equilibrium potentials between the resting potential and that of the declining phase of the 50-msec pulse. Furthermore, the delayed conductance change must have been coupled with the early increase in Ca conductance since the after-potentials were seen only after an initial transient (Fig. 4C and Fig. 6, *left column*).

SUPPRESSION OF BOTH MEMBRANE CA ACTIVATION AND CONTRACTION BY HYPERTONIC SOLUTIONS

To study the kinetics of both the initial transient and the induced spikes under conditions that reduce the risk of damaging microelectrode-impaled fibers, experiments were carried out on fibers bathed in hypertonic solutions (Suarez-Kurtz and Sorenson, 1974). This approach was based on the knowledge that the contraction of frog fast fibers can be eliminated without affecting the kinetics of a Na action potential by soaking the fibers in a hypertonic medium (Hodgkin and Horowicz, 1957).

In the two experiments (Fig. 5) crab fast fibers were bathed in solutions made hypertonic by the addition of Tris propionate. Tris, rather than Na, was used to keep the Na/Ca ratio constant, an important factor in determining membrane Ca currents in crustacean fibers (Orentlicher and Ornstein, 1971). Propionate was chosen because the membrane of the crab fibers is impermeable to this anion. The left-hand column of recordings of current, voltage, and tension in Fig. 5 were obtained from a fiber bathed first in control saline (*top set*), and then in solutions of 1.45 (*middle set*) and 1.7 (*bottom set*) times the control tonicity. When the fiber was bathed in hypertonic media, the threshold for the initial transient changed, as did the amplitude and time course. As anticipated, when the fiber was exposed to the solutions of higher tonicity, the tension was blocked.

A second fiber, which was continuously exposed to 10 mM Ba to convert the graded activity to all-or-none Ca spikes, was bathed in three solutions

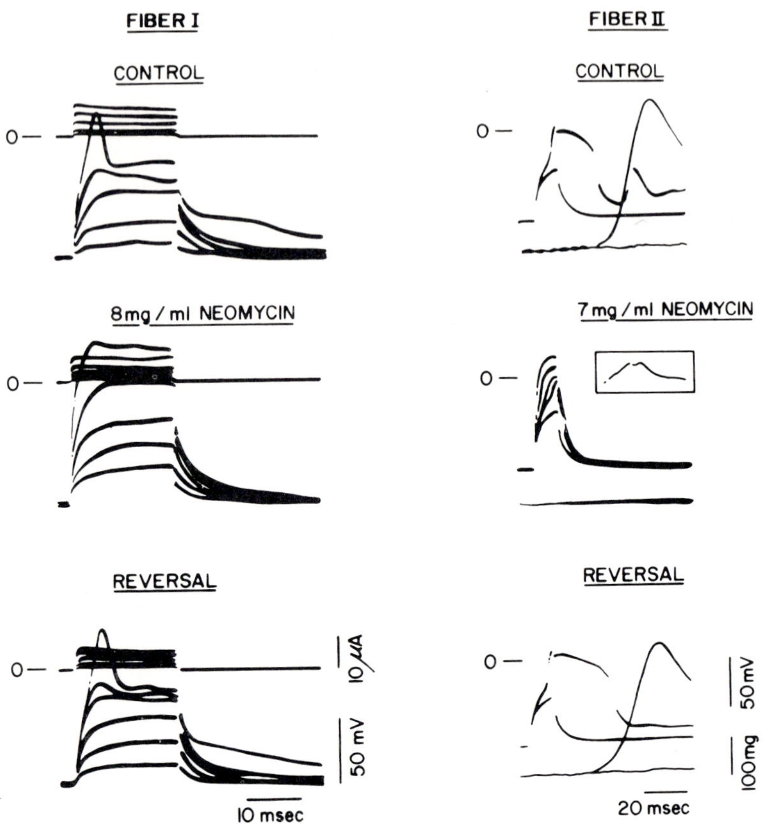

FIG. 6. Inhibition of Ca activation and tension by neomycin. *Fiber I:* Inhibition of membrane Ca activation by neomycin. *Fiber II:* Blockade of procaine-induced spikes by neomycin. Further description in text. (From Suarez-Kurtz, 1974).

of decreasing tonicity (Fig. 5, *right-hand column*). Both the Ca spike and the tension were absent while the fiber was bathed in a solution of 1.9 times the control tonicity and depolarized to the zero level. Upon reduction of the tonicity to 1.5 times the control value, a slow-graded membrane activation appeared, but it was not accompanied by tension. After further reduction of the tonicity to 1.35 of control, the fiber responded to depolarization with an all-or-none spike and a twitch. The action of hypertonic solutions in blocking both the spike and the tension in crab fibers differs from the effect in frog fibers, where only tension is affected. The mechanisms that give rise to Ca action potentials seem to be sensitive to this treatment, whereas those generating Na spikes are insensitive. The findings are another example of the inseparability of membrane Ca activation from the development of tension.

THE BLOCKAGE OF MEMBRANE CA ACTION
AND TWITCH TENSION BY THE AMINOGLYCOSIDE
ANTIBIOTICS

Because of reports that neomycin and streptomycin depress Ca-dependent signaling processes in other biological systems, their effects on fast crab fibers were examined (Suarez-Kurtz, 1974). These antibiotics inhibit the release of neurohumoral transmitters in autonomic ganglia (Corrado and Ramos, 1958) and at the myoneural junction (Vital-Brazil and Corrado, 1957; Elmqvist and Josefsson, 1962; Meiri and Rahamimoff, 1972). The release of neurohormones is adversely affected by these antibiotics (Batalla, 1974) as is ECC in smooth muscle preparations (Pimenta de Morais, Corrado, and Suarez-Kurtz, 1974).

In the fast crab fibers, neomycin inhibits the initial transient (Fig. 6, *left column*), the Ca spike, and the twitch tension (Fig. 6, *right column*). At a concentration of 8 mg/ml neomycin, an initial transient could not be evoked even when the membrane potential was reversed to 20 mV inside positive (Fig. 6, *left column, middle tracings*). The effects of neomycin were promptly reversed by washing the fiber with control saline (Fig. 6, *bottom tracings*). Neomycin (7 mg/ml) was applied to another fiber that responded with Ca spikes and twitches to suprathreshold depolarizations due to treatment with procaine (1 mg/ml). The spikes are shown before (Fig. 6, *right column, top tracings*) and after (*bottom tracings*) exposing the fiber to neomycin. The voltage microelectrode was partially dislodged by the twitch tension when the fiber was bathed in control saline. Maintaining the impalement of the microelectrode after exposing the fiber to neomycin was not a problem because tension was blocked (Fig. 6, *right column, middle tracings*). However, a slow and attenuated graded response was observed for intermediate levels of depolarization. This low level of Ca activation that persisted in the presence of neomycin was more evident when the recording microelectrode was moved several mm toward the tendon end of the fiber (Fig. 6, *inset above middle tracings, right column*). The graded activation outlasted the pulse duration under this recording condition. At the recording sensitivity used to register the twitch tension, a small tension associated with the residual Ca activation could not be recorded. Upon washing out the neomycin with saline containing procaine (1 mg/ml), the Ca spike and twitch tension were restored (Fig. 6, *right column, bottom tracings*).

The inhibitory effect of neomycin on the graded Ca activation, the Ca spike, and the tension add to the evidence supporting our earlier conclusions (Reuben et al., 1967; Suarez-Kurtz et al., 1972) that an inward Ca current serves as the transmembrane signal in ECC of peripheral striated muscle fibers.

SUMMARY

Since the dispute concerning a direct or indirect involvement of external Ca in the excitation process in frog twitch fibers has not been resolved (see Stefani and Chiarandini, 1973), we should perhaps restrict the applicability of the hypothesis to ECC in crustacean fibers. We believe, however, that the reported action of Mn on frog twitch fibers (Chiarandini and Stefani, 1973) is strong evidence of a similar role of an inward Ca current. The disagreement among researchers studying the ECC process in frog fibers is based mainly upon measurements of the fiber's ability to contract while bathed in a so-called Ca-free solution. Both an attenuation of tension (Stefani and Chiarandini, 1973) and no effect on tension (Armstrong, Bezanilla, and Horowicz, 1972) in "Ca-free" solutions have been reported. The different results appear to be due to the different methods of preparing such a solution (see discussion section of Stefani and Chiarandini, 1973).

In this report, as in earlier studies (Reuben et al., 1967; Suarez-Kurtz et al., 1972), we circumvented the problems associated with trying to reduce the amount of Ca available at the surface of a fiber for transporting inward current. Instead, we enhanced or blocked Ca fluxes by experimental agents and correlated membrane Ca conductance with tension and its initiation.

Our evidence provides both additional support and comparative data to bolster the correlation between membrane Ca activation and development of tension. The comparative data suggest to us that rapid Ca activation in fast crab fibers is necessary for initiation of the twitch-like tensions and that prolonged and slow Ca activation in crayfish muscle accounts for the slow development and prolonged tensions in those fibers.

ACKNOWLEDGMENTS

The work in G.S.-K.'s laboratory is supported in part by grants from the Brazilian National Research Council (CNPq) and the Council for Graduate Education of the Federal University of Rio de Janeiro.

J.P.R.'s laboratory is supported in part by funds from the Muscular Dystrophy Associations of America, Inc.; by Public Health Service research grants NS 03728 and HL 16082; training grant NS 05328 and grant NS 11766 from the National Institute of Neurological Diseases and Stroke; and National Science Foundation grant GB 31807X.

REFERENCES

Armstrong, C. M., Bezanilla, F. M., and Horowicz, P. (1972): Twitches in the presence of ethylene glycol bis (β-aminoethyl ether) N, N'-tetraacetic acid. *Biochim. Biophys. Acta* 267:605–608.

Batalla, L. (1974): Efectos de la neomicina en la liberacion de vasopresina y movimientos de radiocalcio en la neurohipofisis de rata. *MSc Thesis, Instituto de Biofisica, Rio de Janeiro.*

Bianchi, C. P. (1968): *Cell Calcium.* Butterworth & Co., London.

Bianchi, C. P. (1969): Pharmacology of excitation-contraction coupling in muscle. *Fed. Proc.* 28:1624–1627.

Bianchi, C. P., and Bolton, T. C. (1967): Action of local anesthetics on coupling systems in muscle. *J. Pharmacol. Exp. Ther.* 157:388–405.

Chiarandini, D. J., Reuben, J. P., Brandt, P. W., and Grundfest, H. (1970): Effects of caffeine on crayfish muscle fibers. I. Activation of contraction and induction of Ca-spike electrogenesis. *J. Gen. Physiol.* 55:640–664.

Chiarandini, D. J., and Stefani, E. (1973): Effects of manganese on the electrical and mechanical properties of frog skeletal muscle fibres. *J. Physiol.* 232:129–147.

Corrado, A. P., and Ramos, A. O. (1958): Neomycin—its curariform and ganglioplegic actions. *Rev. Bras. Biol.* 18:81–85.

Edman, K. A. P., and Grieve, D. W. (1964): On the role of calcium in the excitation-contraction process of frog sartorius muscle. *J. Physiol.* 170:138–152.

Elmqvist, D., and Josefsson, J. D. (1962): Cation requirements of the neuromuscular block produced by neomycin. *Acta Physiol. Scand.* 54:105–110.

Endo, M., Tanaka, M., and Ogawa, Y. (1970): Calcium induced release of calcium from the sarcoplasmic reticulum of skinned skeletal muscle fibres. *Nature* 228:34–36.

Fatt, P., and Ginsborg, B. L. (1958): The ionic requirements for the production of action potentials in crustacean muscle fibres. *J. Physiol.* 142:516–543.

Ford, L. E., and Podolsky, R. J. (1972): Intracellular calcium movements in skinned muscle fibres. *J. Physiol.* 223:21–33.

Girardier, L., Reuben, J. P., Brandt, P. W., and Grundfest, H. (1963): Evidence for anion permselective membrane in crayfish muscle fibers and its possible role in excitation-contraction coupling. *J. Gen. Physiol.* 47:189–214.

Heilbrun, L. V., and Wiercinski, F. J. L. (1947): The action of various cations on muscle protoplasm. *J. Cell. Comp. Physiol.* 29:15–32.

Hodgkin, A. L., and Horowicz, P. (1957): The differential action of hypertonic solution on the twitch and action potential of the muscle fibre. *J. Physiol.* 136:17P.

Ishiko, N., and Sato, M. (1957): The effect of calcium ion on electrical properties of striated muscle fibers. *Jap. J. Physiol.* 7:51–63.

Jenden, D. J., and Reger, J. F. (1963): The role of resting potential changes in the contractile failure of frog sartorius muscles during calcium deprivation. *J. Physiol.* 169:889–901.

Katz, B., and Miledi, R. (1969): Tetrodotoxin-resistant electrical activity in presynaptic terminals. *J. Physiol.* 203:459–487.

Kusano, K. (1968): Further study of the relationship between pre- and postsynaptic potentials in the squid giant synapse. *J. Gen. Physiol.* 52:326–345.

Meiri, U., and Rahamimoff, R. (1972): Neuromuscular transmission: inhibition by manganese ions. *Science* 176:308–309.

Miledi, R., and Slater, C. R. (1966): The action of calcium on neuronal synapses in the squid. *J. Physiol.* 184:473–498.

Orentlicher, M., and Ornstein, R. S. (1971): Influence of external cations on caffeine-induced tension: Calcium extrusion in crayfish muscle. *J. Membr. Biol.* 5:319–333.

Pimenta de Morais, I., Corrado, A. P., and Suarez-Kurtz, G. (1974): Competitive antagonism between calcium and aminoglycoside antibiotics on guinea pig intestinal smooth muscle. *J. Pharmacol. Exp. Ther.* (*submitted for publication*).

Portzehl, H., Caldwell, P. C., and Ruegg, J. C. (1964): The dependence of contraction and relaxation of muscle fibres from the crab *Maia squinado* on the internal concentration of free calcium ions. *Biochim. Biophys. Acta* 79:581–591.

Rasmussen, H. (1970): Cell communication, calcium ion and cyclic adenosine monophosphate. *Science* 170:404–412.

Reuben, J. P., Brandt, P. W., Garcia, H., and Grundfest, H. (1967): Excitation-contraction coupling in crayfish. *Am. Zoologist,* 7:623–645.

Reuben, J. P., Brandt, P. W., and Grundfest, H. (1974): Regulation of myoplasmic calcium concentration in intact crayfish muscle fibers. *J. Mechanochem. Cell Motility,* 2:269–285.

Reuben, J. P., Katz, G. M., and Berman, M. (1969): Two phases of contractile activation induced by Ca injections in crayfish muscle fibers. *Fed. Proc.* 28:711.

Sandow, A. (1965): Excitation-contraction coupling in skeletal muscle. *Pharmacol. Rev.* 17:265–320.

Stefani, E., and Chiarandini, D. J. (1973): Skeletal muscle: Dependence of potassium contractures on extracellular calcium. *Pfluegers Arch.* 343:143–150.

Suarez-Kurtz, G. (1974): Inhibition of membrane calcium activation by neomycin and streptomycin in crab muscle fibers. *Pfluegers Arch.* 349:337–349.

Suarez-Kurtz, G., Reuben, J. P., Brandt, P. W., and Grundfest, H. (1972): Membrane calcium activation in excitation-contraction coupling. *J. Gen. Physiol.* 59:676–688.

Suarez-Kurtz, G., and Sorenson, L. A. (1974): Effect of hypertonic solutions on membrane Ca activation in crab muscle fibers. *An. Acad. Bras. Cienc. In press.*

Takeda, K. (1967): Permeability changes associated with the action potential in procaine-treated crayfish abdominal muscle fibers. *J. Gen. Physiol.* 50:1049–1074.

Vital-Brazil, O., and Corrado, A. P. (1957): The curariform action of streptomycin. *J. Pharmacol. Exp. Ther.* 120:452–459.

Concepts of Membranes in Regulation and Excitation,
edited by M. Rocha e Silva and G. Suarez-Kurtz.
Raven Press, New York © 1975.

Some Characteristics of the Excitation-Contraction Coupling Process in Smooth Muscle

Leon Hurwitz

Department of Pharmacology, University of New Mexico, School of Medicine, Albuquerque, New Mexico 87131

INTRODUCTION

The importance of Ca ions for the generation of a mechanical response in smooth muscle (Hurwitz and Suria, 1971), as well as in cardiac (Langer, 1968) and skeletal muscle (Sandow, 1970), is well recognized. Existing evidence supports the contention that an excitatory drug, by interacting with specific receptors in the smooth muscle membrane, directly or indirectly induces or enhances the activation of a Ca-transport system or a Ca-releasing system (Hurwitz and Suria, 1971). Such a system may reside in the plasma membrane or in the membranes of intracellular organelles or in both (Bohr, 1973; Devine, Somlyo, and Somlyo, 1972; Somlyo and Somlyo, 1971; Hurwitz and Joiner, 1970; Van Breemen, Farinas, Gerba, and McNaughton, 1972; Baudouin, Meyer, Fermandjian, and Morgat, 1972; Somlyo, Somlyo, Devine, Peters, and Hall, 1974). An activated Ca-transport system or Ca-releasing system serves to facilitate the flow of Ca ions from the extracellular medium or from an intracellular Ca pool into the cytoplasm of the muscle fiber. Once in the cytoplasm, the divalent ions can react with contractile elements and thereby initiate a mechanical response (Hurwitz and Suria, 1971).

HYPOTHETICAL SCHEME OF THE EXCITATION-CONTRACTION COUPLING PROCESS

The complex formed by the combination of an excitatory drug and the tissue receptors is usually considered to be the product of a rapid, reversible bimolecular reaction that obeys the mass law (Furchgott, 1966; Furchgott and Bursztyn, 1967). There are indications, at least in some smooth muscles, that Ca ions interact with sites in an activated Ca-transport system in a similar manner (Hurwitz and Joiner, 1970; Hurwitz and Joiner, 1969). The interaction between free cytoplasmic Ca ions and contractile elements also appears to be a rapid, reversible process. Thus a simple scheme depicting the hypothetical manner in which the drug-receptor interaction, the Ca-

55

transport site interaction, the cytoplasmic Ca–contractile protein interaction and the ultimate mechanical response are interconnected may be written as follows (Fig. 1):

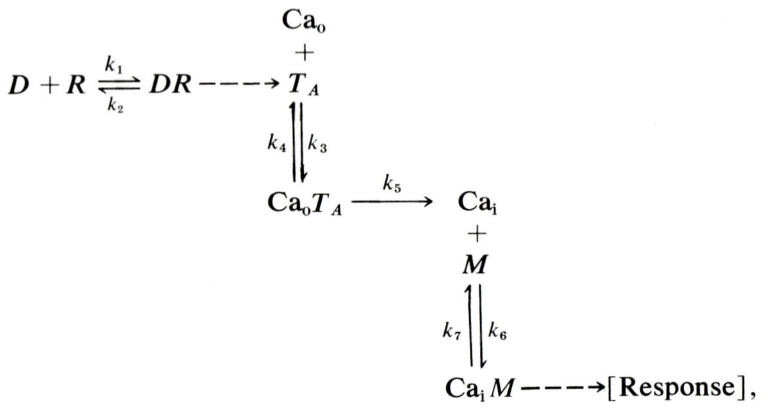

$$
\begin{array}{ccccc}
 & & & \mathrm{Ca_o} & \\
 & & & + & \\
D + R & \underset{k_2}{\overset{k_1}{\rightleftharpoons}} & DR & \dashrightarrow T_A & \\
 & & & \Big\updownarrow k_4 \big\| k_3 & \\
 & & \mathrm{Ca_o}T_A & \xrightarrow{k_5} & \mathrm{Ca_i} \\
 & & & & + \\
 & & & & M \\
 & & & & \Big\updownarrow k_7 \big\| k_6 \\
 & & & & \mathrm{Ca_i}M \dashrightarrow [\text{Response}],
\end{array}
$$

FIG. 1

where D represents the concentration of an excitatory drug in the external environment of the smooth muscle fibers; R is the concentration of free tissue receptors, DR is the concentration of the drug-receptor complex formed; $\mathrm{Ca_o}$ is the concentration of Ca ions in some extracellular or intracellular Ca depot; T_A is the concentration of free sites in an activated Ca-transport system that can combine with the divalent ions; $\mathrm{Ca_o}T_A$ is the concentration of sites occupied by Ca ions in the activated transport system; $\mathrm{Ca_i}$ is the concentration of free cytoplasmic Ca ions; M is the concentration of free sites in the contractile apparatus that can react with Ca ions; $\mathrm{Ca_i}M$ equals the concentration of sites occupied by Ca ions in the contractile apparatus; [Response] refers to the magnitude of the smooth muscle contraction; k_1, k_2, k_3, k_4, k_6, k_7 are reaction rate constants; k_5 is a rate constant that partially determines the velocity at which the Ca-transport site complex moves the divalent ions from an extracellular or intracellular Ca depot to the cytoplasm; and the dashed arrows refer to an assumed complex series of cellular reactions and processes.

Two additional points should be made with respect to the postulated scheme. First, since the major concern here is the sequence of cellular events leading to a smooth muscle contraction, the scheme does not include those cellular processes that regulate the concentration of cytoplasmic Ca ions by actively removing the divalent ions from the cytoplasm (Hurwitz and Suria, 1971; Hurwitz, Fitzpatrick, Debbas, and Landon, 1973; Fitzpatrick, Landon, Debbas, and Hurwitz, 1972; Baudouin et al., 1972). Second, although previous work has indicated that the Ca-transport system exhibits the behavior of a saturable system (Hurwitz and Suria, 1971; Hurwitz and Joiner, 1970; Hurwitz and Joiner, 1969), no definitive statement can be made concerning its basic characteristics. It may consist of a finite number of sites on which Ca must be absorbed before entering membrane pores or a finite number of carrier molecules that ferry the Ca ions to the cytoplasm or some other cellular process.

DRUG-RECEPTOR INTERACTION AND ACTIVATION OF THE CALCIUM TRANSPORT SYSTEM

Further consideration calls attention to a number of other aspects of a drug-induced smooth muscle contraction in need of clarification. One obvious problem is the association between the formation of the drug-receptor complex and the activation of the Ca-transport or Ca-releasing system. If the formation of the drug-receptor complex does indeed induce a cellular reaction or activate a cellular process that facilitates the flow of Ca ions into the cytoplasm, how is this reaction or process linked to drug-filled receptors? Does the occupation of one receptor by an excitatory drug result in the formation or generation of one activated transport site, or are the drug-filled receptors and the activated transport sites interconnected in a more complex fashion?

As an initial attempt to gain information on this point, an investigation was carried out to determine the relationship between the number of tissue receptors occupied by drug molecules and the level of activation of the Ca-transport system (Hurwitz, Hubbard, and Little, 1972). Our experimental approach was similar to that formulated by Furchgott and co-workers for the purpose of determining dissociation constants for agonist-receptor interactions (Furchgott, 1966; Furchgott and Bursztyn, 1967). These workers accepted the view that an excitatory agent reacts with tissue receptors in a rapid reversible manner to form the drug-receptor complex. Each receptor occupied by the agonist evokes a quantum of excitation that was labeled the stimulus. The proportionality constant equating the number of drug-filled receptors to the magnitude of the stimulus was defined as the intrinsic efficacy of the agonist. It denotes the relative intensity with which a drug, by interacting with its receptors, will evoke the stimulus. The ultimate mechanical response of the muscle fiber is presumed to be generated by the stimulus, although its magnitude is not necessarily a linear function of the stimulus. These concepts may be expressed as follows:

$$[\text{Response}] = f(S) = f(\epsilon\,[DR]) = f\frac{\epsilon\,[R_\text{T}]\,[D]}{K_\text{DR} + [D]} \tag{1}$$

where [Response] refers to the magnitude of the mechanical response, f refers to a real, single-valued function that is not necessarily linear, (S) is the magnitude of the stimulus, ϵ is the intrinsic efficacy, $[DR]$ is the concentration of the drug-receptor complex, $[R_T]$ refers to the total concentration of receptors in the tissue, $[D]$ is the concentration of the agonist, and K_{DR} is the dissociation constant for the drug-receptor complex.

It is apparent that any experimental condition that reduces the ϵ or $[R_T]$ term by some fraction, $(1-q)$, will reduce the magnitude of the mechanical response. If, however, the original response can be reinstated by increasing the concentration of agonist, then:

$$f\left(\frac{q\,\epsilon\,[R_T]\,[D']}{K_{DR}+[D']}\right)=f\left(\frac{\epsilon\,[R_T]\,[D]}{K_{DR}+[D]}\right) \tag{2}$$

where $[D']$ is the concentration of agonist that, in the experimental situation, elicits a response equal to that elicited by $[D]$ under control conditions. Solving for $1/[D]$ gives the following equation:

$$\frac{1}{[D]}=\frac{1}{q}\frac{1}{[D']}+\frac{1-q}{qK_{DR}} \tag{3}$$

The straight line defined by Eq. (3) may be determined by taking note of the concentrations of an agonist that will induce several different magnitudes of contraction under control conditions and the concentrations of the agonist that will induce the same group of responses under experimental conditions in which ϵ or $[R_T]$ has been modified. By plotting $1/[D]$ against $1/[D']$ for equal responses, one obtains a straight line with a slope equal to $(1/q)$ and an intercept equal to $(1-q/qK_{DR})$. Therefore:

$$K_{DR}=\frac{\text{Slope}-1}{\text{Intercept}} \tag{4}$$

Furchgott and co-workers (Furchott, 1966; Furchgott and Bursztyn, 1967; Besse and Furchgott, 1967), as well as other investigators (Sastry and Cheng, 1972; Rico, 1971), have employed the poorly reversible antagonist dibenamine to reduce the total number of active receptors, $[R_T]$, in a tissue preparation to some lesser quantity, $q[R_T]$. This procedure has allowed them to apply the experimental approach described above to determine the dissociation constants for a variety of agonist-receptor complexes.

One can show that the existence of a close association between tissue receptors and the Ca-transport system provides still another means of determining the correct value of K_{DR}. The experimental approach in this case is similar to the one described above, but a modification in the Ca ion concentration of the bathing medium is substituted for the introduction of dibenamine. The validity of the procedure is based on the assumption that the degree of activation of the Ca-transport system is directly proportional to the number of tissue receptors occupied by an agonist. If so, then:

$$T_A=\epsilon\,[DR] \tag{5}$$

where T_A is the concentration of activated Ca carriers or widened membrane pores or some other cellular component representing an activated portion of a Ca-transport system, ϵ is the intrinsic efficacy of the agonist, and $[DR]$ is the concentration of the drug-receptor complex. Moreover, if the scheme in Fig. 1 constitutes a reasonably true representation of the cellular events leading to a smooth muscle contraction, then the mechanical response of the muscle may be considered to be some function of the rate of movement of Ca ions into the cytoplasm. The rate of movement of Ca into the cytoplasm, in turn, is dependent on the degree to which the Ca-transport system has been activated and on the degree to which the activated transport system has been saturated with mobilizable ions from the Ca depots. The latter factor can be controlled by adjusting the concentration of Ca ions in the external bathing medium thus:

$$[\text{Response}] = f(k_s T_A), \tag{6}$$

where k_s is a constant that denotes the level of saturation of the Ca-transport system and the other symbols represent quantities previously defined. Since there is an assumed proportionality between the concentration of drug-receptor complex and the degree of activation of the Ca-transport system,

$$[\text{Response}] = f(k_s \, \epsilon \, [DR]) = f\left(\frac{k_s \, \epsilon \, [R_T] \, [D]}{K_{DR} + [D]}\right) \tag{7}$$

When a given response is elicited in the presence of two different extracellular Ca ion concentrations, the relationship between the two levels of agonist ($[D]$ and $[D']$) used to obtain that response is:

$$f\left(\frac{k_s \, \epsilon \, [R_T] \, [D]}{K_{DR} + [D]}\right) = f\left(\frac{k_s' \, \epsilon \, [R_T] \, [D']}{K_{DR} + [D']}\right) \tag{8}$$

where k_s and k_s' represent two different levels of saturation of the activated Ca-transport system. Solving for $1/[D]$ gives:

$$\frac{1}{[D]} = \left(\frac{k_s}{k_s'}\right) \frac{1}{[D']} + \frac{1}{K_{DR}} \left(\frac{k_s}{k_s'} - 1\right) \tag{9}$$

It may be seen that the expression (Slope $-$ 1/Intercept) discloses the true value of the dissociation constant K_{DR} for the drug-receptor complex.

In order to determine whether experiments based on the assumptions and procedure described would produce the anticipated results, an investigation was carried out on isolated longitudinal smooth muscle from the guinea pig ileum and on isolated thoracic aorta from the rabbit. Dose-response curves were obtained with each smooth muscle preparation in the presence of several different extracellular Ca ion concentrations. Since the ultimate purpose in changing the extracellular Ca ion concentration was to modify Ca levels in the fiber depots, the steps taken to modify Ca levels in rabbit aorta were somewhat different from those taken to modify Ca levels in guinea pig in-

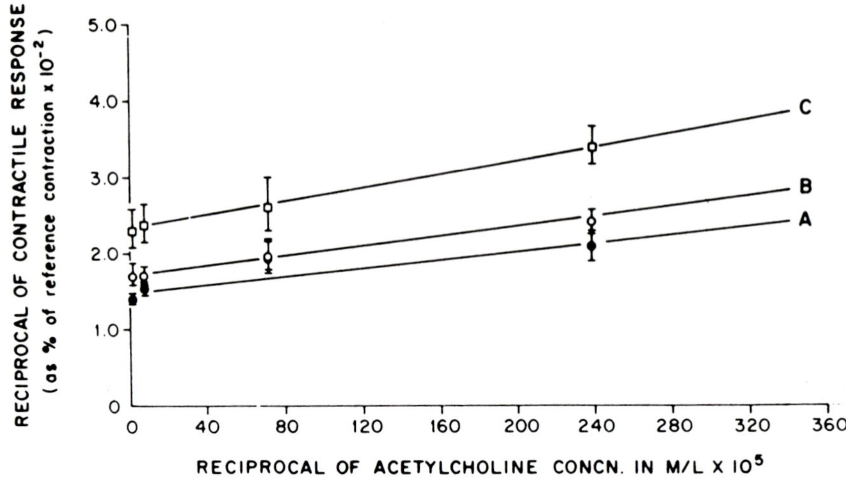

FIG. 2. Reciprocal plot of the dose-response relationship between ACh concentration and sustained isotonic contraction of longitudinal muscle from guinea pig ileum. *A*, bathing medium contained 1.8mM $CaCl_2$; *B*, bathing medium contained 0.9mM $CaCl_2$; *C*, bathing medium contained 0.36mM $CaCl_2$. (Reproduced by permission, from Hurwitz et al., 1972.)

testinal muscle. This was necessary because a portion of the mobilizable Ca pool in rabbit aorta is less readily accessible to the external medium than is the mobilizable Ca pool in longitudinal muscle of guinea pig ileum (Devine et al., 1972; Hurwitz and Joiner, 1970; Hurwitz et al., 1973; Hudgins and Weiss, 1968). Acetylcholine (ACh) was used to induce the mechanical response in guinea pig longitudinal muscle, and norepinephrine (NE) was used to elicit the response in rabbit aortic muscle.

The data in Fig. 2 illustrate the type of experimental results obtained. The figure contains the reciprocal plots of three hyperbolic dose-response curves, representing the magnitudes of the sustained (tonic) isotonic contractions of the longitudinal muscle induced by various concentrations of ACh. Each dose-response curve was obtained in the presence of a different extracellular Ca ion concentration.

Plots denoting the relationship between $1/[D]$ and $1/[D']$ for equal responses are shown in Fig. 3. The figure contains two such straight-line plots. Line A was constructed by matching ACh-induced contractions elicited in the presence of 1.8 mM $CaCl_2$ (control, $1/[D]$) with ACh-induced contractions elicited in the presence of 0.9 mM $CaCl_2$ (experimental, $1/[D']$). Line B was constructed by matching the contractions elicited in the presence of 1.8 mM $CaCl_2$ (control, $1/[D]$) with the contractions elicited in the presence of 0.36 mM $CaCl_2$ (experimental, $1/[D']$). The values of K_{DR} for the ACh-receptor complex calculated in each case from the (Slope − 1/Intercept) were 3.6×10^{-8} M and 2.0×10^{-8} M, respectively.

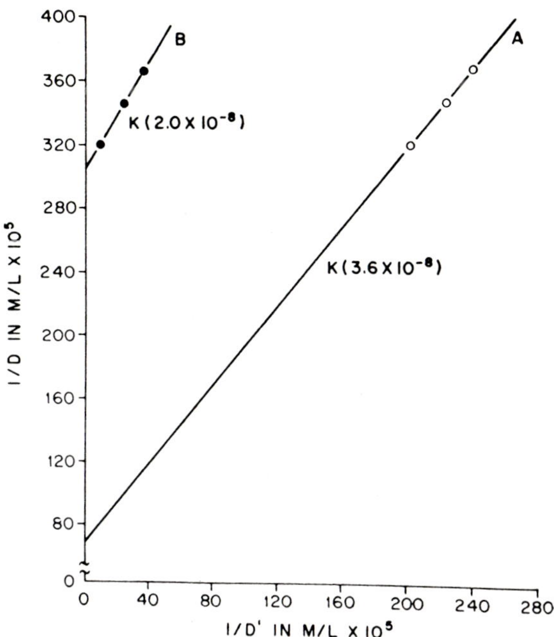

FIG. 3. A $1/[D]$ vs. $1/[D']$ plot for the sustained isotonic contraction of the longitudinal muscle of the guinea pig ileum. See text for explanation. (Reproduced by permission, from Hurwitz et al., 1972.)

Two additional groups of experiments, similar in design, were carried out to determine the values of K_{DR} from measurements of the peak (phasic) isotonic contractions of guinea pig longitudinal muscle induced by ACh and measurements of the isometric contractions of rabbit aorta induced by NE. In the former case, the K_{DR} values for the ACh-receptor complex in guinea pig longitudinal muscle were found to be 1.7×10^{-8} M and 5.4×10^{-8} M. In the latter, the K_{DR} values for the NE-receptor complex in the aortic muscle were found to be 0.64×10^{-7} M and 1.6×10^{-7} M.

A comparison of the magnitudes of the dissociation constants determined by modifying the Ca ion concentration with the magnitudes of the dissociation constants determined by introducing dibenamine revealed a large discrepancy in one case but not in the other. As indicated above, the K_{DR} for the ACh-receptor complex in guinea pig longitudinal muscle obtained by modifying the Ca concentration varied from 1.7×10^{-8} to 5.4×10^{-8} M; whereas the value obtained by using dibenamine has been reported to be 1.0×10^{-6} M to 2.6×10^{-6} M (Sastry and Cheng, 1972; Rico, 1971). The difference in the results seen with each of the two procedures is quite striking, extending between one and two orders of magnitude. On the other hand, the magnitudes of the K_{DR} for the NE-receptor complex in rabbit aortic muscle determined by each of the two procedures differed to a much lesser degree. The magnitude of the K_{DR} found by modifying the Ca ion concentration varied from 0.64×10^{-7} to 1.6×10^{-7} M; whereas the magnitude found by employing dibenamine has been reported to be about 1.8×10^{-7} M (Besse and Furchgott, 1967).

Since dissociation constants cannot be correctly determined by modifying the calcium concentration unless a linear relationship exists between the number of drug-receptor complexes formed and the level of activation of the Ca-transport system, the low value of the K_{DR} found in experiments with the longitudinal muscle of guinea pig ileum strongly indicates that such a relationship does not exist in this smooth muscle. A low value for the K_{DR} of the ACh-receptor interaction may, however, be expected if the Ca-transport system develops a high level of activation when only a small fraction of the total cholinergic receptor population is occupied by ACh. By contrast, the value of K_{DR} for the NE-receptor complex is similar to that found by introducing dibenamine. This would suggest that the level of activation of the Ca-transport system is a linear or nearly linear function of the number of norepinephrine-filled receptors.

Other investigators working with these two test systems have reached analogous conclusions regarding the relationship between the contractile response and the drug-receptor complex. In guinea pig longitudinal muscle, a near-maximum contraction can be generated when, supposedly, only a small number of the total cholinergic receptors are occupied by ACh (Sastry and Cheng, 1972). In rabbit aorta, the contractile response appears to be more nearly proportional to the fraction of the total adrenergic receptors occupied by NE (Furchgott, 1966; Taylor and Green, 1970). Thus the longitudinal muscle is said to have a large number of spare receptors; whereas the aortic muscle seems to contain fewer spare receptors.

The concept of spare receptors was introduced, initially, to explain the surprising results obtained with such poorly reversible blocking agents as dibenamine (Nickerson, 1956, 1957; Furchgott, 1955). Although sufficiently high concentrations of this drug can lower the maximum of a dose-response curve, there is a range of concentrations of dibenamine which, in some cases, appears to inhibit the mechanical response in a purely competitive fashion (Furchgott, 1955; Nickerson, 1957). These experimental observations become understandable if one accepts the view that a maximum or near-maximum response may be elicited when only a small fraction of the total receptor population of the muscle is occupied by agonist molecules (Nickerson, 1956, 1957; Furchgott, 1955; Stephenson, 1956). Accordingly, if a partial quasi-irreversible inactivation of receptors by relatively small concentrations of dibenamine leaves unblocked more than the critical number of receptors needed to produce a maximum response, the ultimate effect of the inhibitory agent would simply be a shift in the dose-response curve (response versus log dose) to the right. Higher concentrations of dibenamine that reduce the active receptor population to less than this critical number would also reduce the magnitude of the maximum response.

On the other hand, there is the alternative possibility that relatively small concentrations of dibenamine may interact in a poorly reversible manner with some allosteric site on the receptor and that this interaction may somehow serve to decrease the affinity of some specific agonist for the receptor. Furthermore, the reduction in the maximum response induced by higher concentrations of dibenamine may be a consequence of the interaction of the inhibitory agent with some other cellular sites involved in the contractile process. If this were so, there would be no need to invoke the concept of the spare receptor. Moreover, it would be a mistake to interpret the significance of the K_{DR} value obtained by modifying the calcium ion concentration by comparing it with the K_{DR} value obtained by introducing dibenamine.

In an effort to resolve this question a group of experiments was carried out on the longitudinal muscle of guinea pig ileum. The experiments were designed to determine whether the saturation effect that a dose-response curve exhibits is due primarily to the progressive saturation of tissue receptors or to the progressive saturation of other cellular sites or processes involved in contraction. As a first step in the experimental procedure, a contractile response was induced in the ileal muscle by introducing 2×10^{-8} M ACh. A second response of equal magnitude was then induced by introducing—usually after several trials—an appropriate concentration of histamine. This concentration of histamine, together with a concentration of 2×10^{-8} M ACh, was added simultaneously to the bathing medium to produce a third contractile response. Finally, a fourth response was elicited by introducing twice the original concentration of ACh, namely 4×10^{-8} M.

The records in Fig. 4 illustrate the results obtained in one representative experiment. It is apparent in the figure that the magnitude of the smooth muscle contraction produced by 4×10^{-8} M ACh, although it is distinctly a submaximal response, is considerably less than twice the magnitude of the contraction produced by 2×10^{-8} M ACh. This finding is indicative of the saturation effect that occurs in the chain of events leading to a contraction. It may also be seen that the contraction produced by the simultaneous addition of histamine and 2×10^{-8} M ACh is approximately equal in magnitude to the contraction produced by 4×10^{-8} M ACh. Since histamine and ACh interact with two different receptor populations, the fact that the two agonists together do not elicit a substantially greater response than does twice the concentration of ACh supports the argument that the progressive saturation of cholinergic receptors is not a primary factor in determining the configuration of the dose-response curve. It follows, then, that concentrations of ACh up to 4×10^{-8} M occupy only a small fraction of the cholinergic

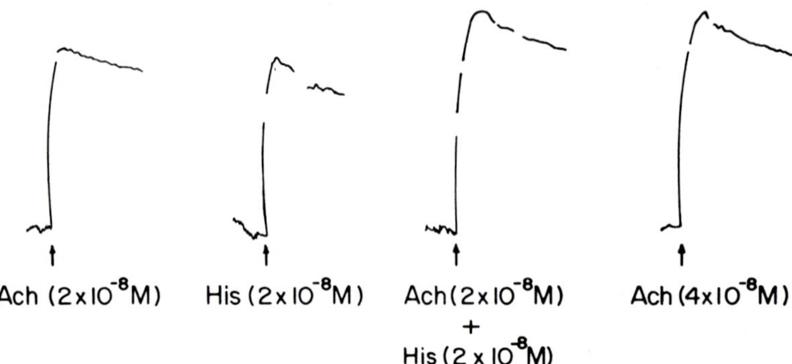

FIG. 4. Drug-induced isotonic contractions of the longitudinal smooth muscle, isolated from guinea pig ileum. His represents histamine.

receptors in the ileal muscle. If this were not the case, one would expect the agonist to occupy a sufficient number of the receptors to produce a saturation effect at this step in the generation of the final response and, as a result, evoke a smaller contraction than that seen in the presence of the combination of ACh and histamine. These simple experiments, thus, favor the conclusion expressed earlier that concentrations of ACh up to 4×10^{-8} M occupy only a small fraction of the cholinergic receptors and, further, that the K_{DR} of the ACh-receptor complex is considerably larger than 4×10^{-8} M.

The significant point to be gained from the studies performed on longitudinal muscle of the guinea pig ileum is that the relationship between the amplitude of the mechanical response and the concentration of ACh does not reflect the saturation of cholinergic receptors. Nor is it determined solely by the limitations of the contractile apparatus. It would appear that a step as early as the transport of Ca into the cytoplasm may reach near-maximum velocity when only a small portion of the receptors is occupied by an agonist. Thus the concept of spare receptors holds for the relationship between the drug-receptor complex and the transport of calcium into the cytoplasm as well as for the relationship between the drug-receptor complex and the final mechanical response.

In view of these findings there seems to be little or no basis for postulating that each receptor occupied by an excitatory agent opens or widens one Ca pore or is converted into one Ca carrier in the smooth muscle membrane. If this were true, one would expect to find a linear relationship between the activity of the Ca-transport or Ca-releasing system and the number of receptors filled with the agonist. As indicated above, this relationship appeared to approach linearity in the aortic smooth muscle but not in the ileal muscle. Accordingly, direct coupling between drug-receptor complexes and Ca carriers or Ca pores on a one-to-one basis does not seem to be the general rule in smooth muscle.

LOCATION OF MOBILIZABLE RESERVOIRS OF CALCIUM

A second question still in need of clarification relates to the location of mobilizable Ca pools in smooth muscle fibers. Any attempt to gain a thorough understanding of the events that lead to a smooth muscle contraction must include a consideration of the locations of the Ca stores from which the divalent ions can be transported into the cytoplasm to activate the contractile machinery. In striated muscle fibers the Ca ions that serve as activators of the contractile machinery are mobilized from a pool of Ca sequestered in the lateral cisternae of the sarcoplasmic reticulum (Zierler, 1974). In smooth muscle fibers the loci from which activator ions are mobilized have not, as yet, been clearly established. Some information has been gleaned from physiological experiments in which the mechanical responses of the muscle fibers were measured in an extracellular environment that was subjected to drastic modifications in its Ca ion concentration. These experiments indicated that essentially two different reservoirs of mobilizable Ca may be present in smooth muscle fibers (Van Breemen and Daniel, 1966; Hiraoka, Yamogishi, and Sano, 1968; Hinke, 1965; Hinke, Wilson, and Burnham, 1964; Daniel, 1965; Jhamandas and Nash, 1967; Hudgins and Weiss, 1968). One is the Ca that is in the extracellular medium or is loosely bound to the muscle fiber. The other is Ca that is tightly bound or sequestered in the fiber.

The degree to which loosely bound or sequestered pools of Ca serve to provide the activator ions needed for contraction differs in different types of smooth muscle fibers. In the smooth muscle of the aorta of the rabbit, for example, both tightly and loosely bound pools of Ca appear to play significant roles in the generation of the mechanical response (Hudgins and Weiss, 1968). By contrast, the generation of a contractile response in the longitudinal muscle of the guinea pig ileum depends primarily on activator ions that arise from a loosely bound pool of Ca (Hurwitz and Joiner, 1970). It has been suggested that one important cellular depot that contains mobilizable calcium is the sarcoplasmic reticulum. This suggestion is based on the observation that a crude correlation exists between the quantities of sarcoplasmic reticulum found in various smooth muscle fibers and their capacities to develop a contractile response in a Ca-free medium (Devine, et al., 1972). Moreover, Somlyo and Somlyo (1971) have shown that strontium, a divalent ion with physiological actions resembling those of Ca, may accumulate in the sarcoplasmic reticulum as well as in the mitochondria of smooth muscle fibers. Other experimental studies have also pointed to the SR, the mitochondria, and, in addition, to the plasma membrane as possible loci from which Ca may move into the cytoplasm to induce a contraction and to which the divalent ions may be transported during relaxation (Bohr, 1973; Somlyo, et al., 1974).

Although previous efforts to localize physiologically significant stores of Ca in smooth muscle have produced highly suggestive results, two questions still remain. First, where do Ca ions concentrate in various types of smooth muscle fibers? Second, what physiological role do these divalent ion stores play in the contraction-relaxation cycle of the smooth muscle fibers? The first of these two questions was investigated by employing a histological method which permitted the visualization of calcium deposits in the muscle fibers (Debbas, Hoffman, Landon, and Hurwitz, 1973, 1974; Debbas, 1973). The method involved the use of potassium pyroantimonate which forms electron-opaque precipitates with calcium ions. The tissue preparation investigated was the thoracic aorta of the rabbit. Komnick (1962) first introduced the use of potassium pyroantimonate to localize Na ions in cells and tissues. However, subsequent work showed that the pyroantimonate ion will also precipitate Ca and Mg ions (Komnick and Komnick, 1963; Klein, Yen, and Klein-T, 1972; Tandler, Libanati, and Sanchis, 1970; Kierszenbaum, Libanati, and Tandler, 1971).

Because of the lack of specificity of the precipitates formed with pyroantimonate ion, the aortic muscle was isolated from the animal and incubated for 50 min in a modified physiological bathing medium in which K ions were substituted for Na ions and Ca and Mg ions were omitted. The tissue was then fixed in a potassium pyroantimonate-osmium tetroxide (OsO_4) fixative and prepared for examination in the electron microscope. This procedure minimized the concentrations of Na, Ca, and Mg ions present in the vascular tissue. Consequently, only sparse amounts of electron-opaque precipitates of pyroantimonate were found in the cytoplasm or in the cytoplasmic organelles of the smooth muscle fiber. If the initial incubation step was not carried out, the aortic smooth muscle was found to contain randomly distributed electron-opaque particles throughout the cytoplasm and in the cytoplasmic organelles. These observations are illustrated in Fig. 5A and B. Figure 5A is an electron micrograph of thoracic aorta isolated from the rabbit, cut into small segments, fixed immediately in a potassium pyroantimonate-OsO_4 fixative, and prepared for examination in the electron microscope. It is obvious that electron-opaque pyroantimonate

FIG. 5. *A:* Aortic smooth muscle fixed immediately in potassium pyroantimonate-OsO_4 without prior incubation. The nucleus (N) and nucleolus (Nu) contain abundant reaction product. Nonspecific reaction product in the cytoplasm (CP) is abundant. Reaction product also occurs in the mitochondria (M) and at the plasma membrane (*arrow*). The sarcoplasmic reticulum (SR) has a slight amount of reaction product. (Magnification 21,000 ×) *B:* Aortic smooth muscle incubated for 50 min at 25°C in the modified physiological bathing medium containing no added $CaCl_2$. It was then fixed in potassium pyroantimonate OsO_4. The nucleus and nucleolus contain appreciable amounts of reaction product. The mitochondria (M) have little or no reaction product, while the sarcoplasmic reticulum (SR) has none. Very little reaction product adheres to inner surface of plasma membrane (*arrow*). (Magnification 22,050 ×) Reproduced with permission from Debbas et al., in press.)

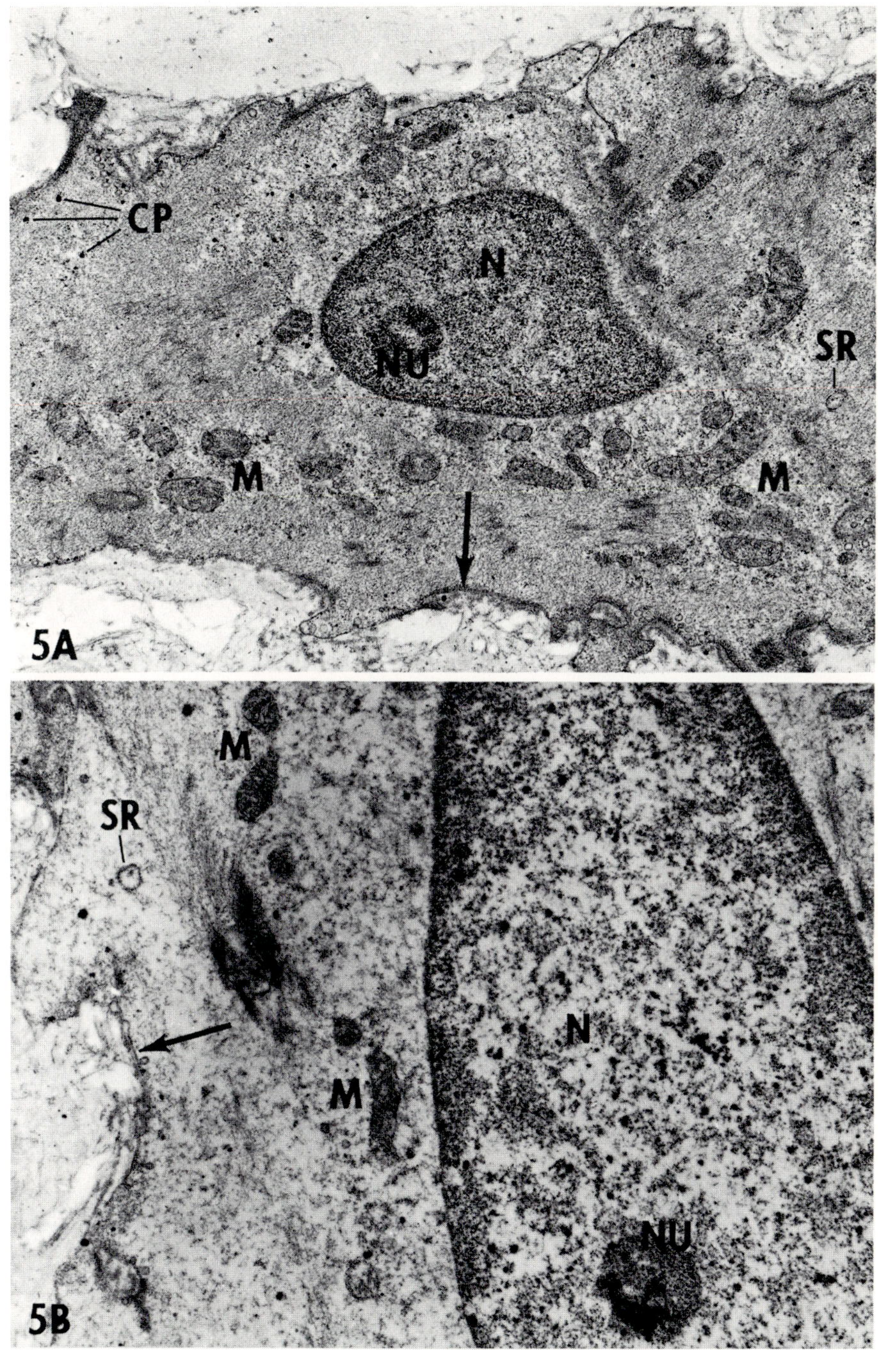

CP
N
NU
SR
M
M
5A
M
SR
M
N
NU
5B

precipitates (reaction product) are present in the cytoplasm, in various cytoplasmic organelles, in the nucleus, and in the nucleolus. This finding is presumed to indicate that there is a random distribution of precipitable inorganic ions, consisting primarily of Na ions. Some of the reaction product may also represent pyroantimonate precipitates of Ca and Mg ions. Figure 5B is an electron micrograph of an aortic muscle that was first incubated for 50 min in the modified high-K bathing solution before it was fixed and prepared for examination. Although an appreciable concentration of pyroantimonate precipitate is still found in the nucleus and nucleolus, very little reaction product can be seen in the cytoplasm or in the cytoplasmic organelles.

Finally, if the isolated thoracic aorta is incubated for 50 min in the modified high-K bathing solution to which 10.8 mM $CaCl_2$ has been added, subsequent fixation of the tissue in the potassium pyroantimonate-OsO_4 fixative reveals discrete areas of concentration of electron-opaque precipitates within the cell. These areas of concentration are illustrated in Fig. 6A and B. The electron micrographs show that only sparse amounts of reaction product appear in the cytoplasm; but appreciable quantities are present in the mitochondria, the sarcoplasmic reticulum and at the inner surface of the plasma membrane. The inference is that the aortic smooth muscle fibers incubated in the K-rich, Ca-rich bathing solution can take up Ca ions from the external medium. The Ca that enters the interior of the cell does not reach a concentration in the cytoplasm high enough to form a detectable precipitate with the pyroantimonate ion, although it is high enough to produce a partial contraction. However, a significant portion of the divalent ion is presumably removed from the bulk of the cytoplasm and becomes highly concentrated in the mitochondria, the sarcoplasmic reticulum and at the inner surface of the plasma membrane. By extrapolation from other studies (Fitzpatrick et al., 1972; Hurwitz et al., 1973; Baudouin et al., 1972), it would appear that these organelles possess ATP-dependent Ca pumps that act to concentrate the divalent ions. By contrast, there was no visible evidence of a concentration of Ca ions occurring at the nuclear membrane.

FIG. 6. *A:* Aortic smooth muscle incubated for 50 min at 25°C in the modified physiological bathing solution containing 10.8 mM $CaCl_2$; then fixed in potassium pyroantimonate-OsO_4. Portions of several smooth muscle cells are shown. Reaction product is abundant on the inner surface of the plasma membrane (*arrows*) as well as within mitochondria (M). Reaction product is also present on profiles of the sarcoplasmic reticulum (SR) and to lesser extent in the nucleus (N). (Magnification, 25,000 ×) *B:* Aortic smooth muscle incubated and fixed as in Fig. 6A. Peripherally located sarcoplasmic reticulum (SR) is dilated and contains dense accumulation of reaction product as does the inner surface of the plasma membrane (*arrows*) of the smooth muscle cell. (Magnification, 35,000 ×) (Reproduced with permission from Debbas et al., in press.)

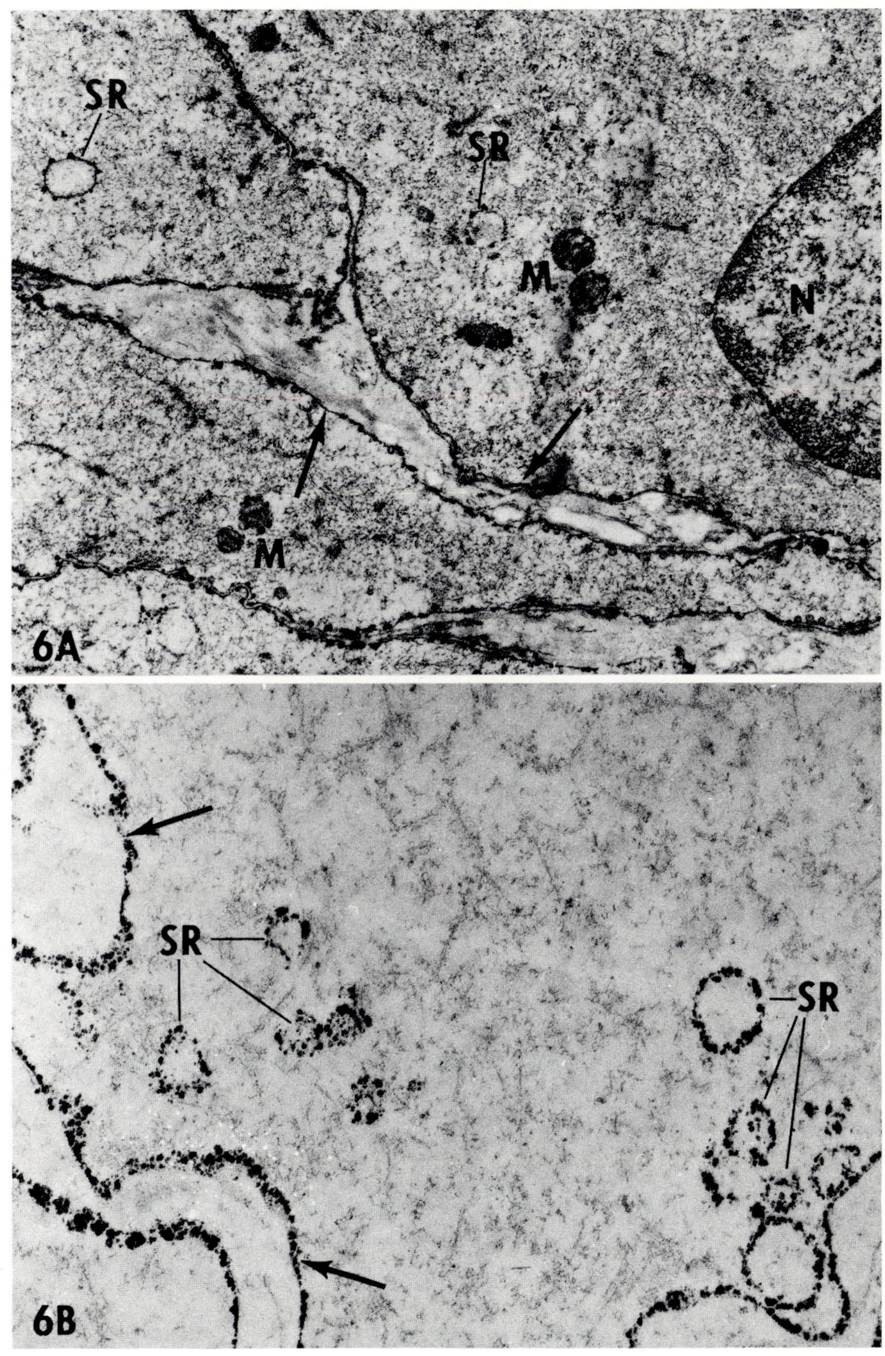

6A

6B

CONCLUDING REMARKS

The experimental studies and results outlined here thus provide some indication of the nature and magnitude of the problem confronting investigators who seek to delineate and relate the series of cellular events that occur when an excitatory drug elicits a contractile response in a smooth muscle fiber. Although there is a general consensus that the interaction between the excitatory drug and specific tissue receptors activates some sort of Ca-transport or Ca-releasing system, experiments performed to ascertain the relationship between these two cellular processes did not support the concept that each drug-filled receptor may be directly associated with one Ca pore or one Ca carrier. We observed, in one case at least, that the occupation of only a small fraction of the total number of receptors by an agonist was sufficient to achieve near-maximal activation of the Ca-transport system. Accordingly, limitations of the latter system, rather than receptor saturation, would be expected to have the greater influence in determining the configuration of a dose-response curve.

With regard to the locations of cellular depots from which Ca ions may be transported into the cytoplasm for contraction, the histological data obtained on the thoracic aorta of the rabbit is more or less consistent with conclusions reached in other investigations. These data implicate the sarcoplasmic reticulum, the inner surface of the plasma membrane, and the mitochondria as possible sources of, and as possible sinks for, the activator ions. One should emphasize, however, that the localization of areas where Ca ions can be concentrated does not by itself establish the roles of these Ca depots in the contraction-relaxation cycle of the smooth muscle fiber. Whether each of the Ca pools identified above serves as a source of activator ions or as a sink for activator ions is still to be determined.

ACKNOWLEDGMENTS

The author is indebted to Dr. Gamil Debbas for his generous help in the preparation of this article. The work was supported by grant HL 16179 from the National Heart and Lung Institute and by grant GB 36059 from the National Science Foundation.

REFERENCES

Baudouin, M., Meyer, P., Fermandjian, S., and Morgat, J. (1972): Calcium release induced by interaction of angiotensin with its receptors in smooth muscle cell microsomes. *Nature* 235:336–338.

Besse, J., and Furchgott, R. F. (1967). Dissociation constants and relative efficacies of agonists acting on adrenergic α-receptors. *Fed. Proc.* 26:401.

Bohr, D. F. (1973): Vascular smooth muscle updated. *Circ. Res.* 32:665–672.

Daniel, E. E. (1965): Attempted synthesis of data regarding divalent ions in muscle function. In: *Muscle,* edited by W. M. Paul et al., pp. 295–313. Pergamon Press, New York.

Debbas, G. (1973): Localization of calcium in vascular smooth muscle by electronmicroscopic cytochemistry. Ph.D. thesis. Department of Pharmacology, Vanderbilt University, Nashville, Tenn.

Debbas, G., Hoffman, L., Landon, E. J., and Hurwitz, L. (1973): Electronmicroscopic localization of calcium in vascular smooth muscle. *Fed. Proc.* 32:767.

Debbas, G., Hoffman, L., Landon, E. J. and Hurwitz, L. (1974): Electronmicroscopic localization of calcium in vascular smooth muscle. *Anat. Rec. In press.*

Devine, C. E., Somlyo, A. V., and Somlyo, A. P. (1972): Sarcoplasmic reticulum and excitation-contraction coupling in mammalian smooth muscles. *J. Cell Biol.* 52:690–718.

Fitzpatrick, D. F., Landon, E. J., Debbas, G., and Hurwitz, L. (1972): A calcium pump in vascular smooth muscle. *Science* 176:305–306.

Furchgott, R. F. (1955): The pharmacology of vascular smooth muscle. *Pharmacol. Rev.* 7:183–265.

Furchgott, R. F. (1966): The use of β-halo-alkylamines in the differentiation of receptors and in the determination of dissociation constants of receptor-agonist complexes. *Adv. Drug Res.* 3:21–56.

Furchgott, R. F., and Bursztyn, P. (1967): Comparison of dissociation constants and of relative efficacies of selected agonists acting on parasympathetic receptors. *Ann. N.Y. Acad. Sci.* 144:882–899.

Hinke, J. A. M. (1965): Calcium requirements for noradrenaline and high potassium ions concentrations in arterial smooth muscle. In: *Muscle,* edited by W. M. Paul et al., pp. 269–284. Pergamon Press, New York.

Hinke, J. A. M., Wilson, M. L., Burnham, S. C. (1964): Calcium and the contractility of arterial smooth muscle. *Am. J. Physiol.* 206:211–217.

Hiraoka, M., Yamogishi, S., Sano, T. (1968): Role of calcium ions in the contraction of vascular smooth muscle. *Am. J. Physiol.* 214:1084–1089.

Hudgins, P. M., and Weiss, G. B. (1968): Differential effects of calcium removal upon vascular smooth muscle contraction induced by norepinephrine, histamine, and potassium. *J. Pharmacol. Exp. Ther.* 159:91–97.

Hurwitz, L., Hubbard, W., and Little, S. (1972): The relationship between the drug-receptor interaction and calcium transport in smooth muscle. *J. Pharmacol. Exp. Ther.* 183:117–125.

Hurwitz, L., Fitzpatrick, D. F., Debbas, G., and Landon, E. J. (1973): Localization of calcium pump activity in smooth muscle. *Science* 179:384–386.

Hurwitz, L., and Joiner, P. D. (1969): Excitation-contraction coupling in smooth muscle. *Fed. Proc.* 28:1629–1633.

Hurwitz, L., and Joiner, P. D. (1970): Mobilization of cellular calcium for contraction in intestinal smooth muscle. *Amer. J. Physiol.* 218:12–19.

Hurwitz, L., and Suria, A. (1971): The link between agonist action and response in smooth muscle. *Annu. Rev. Pharmacol.* 11:303–326.

Jhamandas, K. H., and Nash, C. W. (1967): Effects of inorganic anions on the contractility of vascular smooth muscle. *Can. J. Physiol. Pharmacol.* 45:675–682.

Kierszenbaum, A. L., Libanati, C. M., and Tandler, C. J. (1971): The distribution of inorganic cations in mouse testis. *J. Cell Biol.* 48:314–323.

Klein, R. L., Yen, S. S., and Klein-T, A. (1972): Critique on the K-pyroantimonate method for semiquantitative estimation of cations in conjunction with electron microscopy. *J. Histochem. Cytochem.* 20:65–78.

Komnick, H. (1962): Elektronenmikroskopische lokalisation von Na^+ und Cl^- in zellen und geweben. *Protoplasma* 55:414–418.

Komnick, H., and Komnick, U. (1963): Elektronenmikroskopische untersuchungen zur funktionellen morphologie des ionentransportes in der salzdrüse von larus argentatus. *Z. Zellforsch. Mikrosk. Anat.* 60:163–203.

Langer, G. A. (1968): Ion fluxes in cardiac excitation and contraction and their relation to myocardial contractility. *Physiol. Rev.* 48:708–757.

Nickerson, M. (1956): Receptor occupancy and tissue response. *Nature* 178:697–698.

Nickerson, M. (1957): Nonequilibrium drug antagonism. *Pharmacol. Rev.* 9:246–259.

Rico, J. M. G. T. (1971): The influence of calcium on the activity of full and partial muscarinic agonists. *Eur. J. Pharmacol.* 13:218–229.

Sandow, A. (1970): Skeletal muscle. *Annu. Rev. Physiol.* 32:87–138.

Sastry, B. V. R., and Cheng, H. C. (1972): Dissociation constants of D- and L-lactoylcholines and related compounds at cholinergic receptors. *J. Pharmacol. Exp. Ther.* 180:326–339.

Somlyo, A. P., Somlyo, A. V., Devine, C. E., Peters, P. D., and Hall, T. A. (1974): Electron microscopy and electron probe analysis of mitochondrial cation accumulation in smooth muscle. *J. Cell Biol.* 61:723–742.

Somlyo, A. V., and Somlyo A. P. (1971): Strontium accumulation by sarcoplasmic reticulum and mitochondria in vascular smooth muscle. *Science* 174:955–958.

Stephenson, R. P. (1956): A modification of receptor theory. *Br. J. Pharmacol.* 11:379–393.

Tandler, C. J., Libanati, C. M., and Sanchis, C. A. (1970): The intracellular localization of inorganic cations with potassium pyroantimonate. *J. Cell Biol.* 45:355–366.

Taylor, J., and Green, R. D. (1970): Evidence for an antagonistic action of diazoxide at alpha-adrenergic receptors in rabbit aorta. *Eur. J. Pharmacol.* 12:385–387.

Van Breemen, C., and Daniel, E. E. (1966): The influence of high potassium depolarization and acetylcholine on calcium exchange in the rat uterus. *J. Gen. Physiol.* 49:1299–1317.

Van Breemen, C., Farinas, B. R., Gerba, P., and McNaughton, E. D. (1972): Excitation-contraction coupling in rabbit aorta studied by the lanthanum method for measuring cellular calcium influx. *Circ. Res.* 30:44–54.

Zierler, K. L. (1974): Mechanism of muscle contraction and its energetics. In: *Medical Physiology,* edited by V. B. Mountcastle, pp. 77–120. Mosby, St. Louis.

Concepts of Membranes in Regulation and Excitation,
edited by M. Rocha e Silva and G. Suarez-Kurtz.
Raven Press, New York © 1975.

Calcium Translocation in Rat Vas Deferens in the Light of Receptor Theory

A. Jurkiewicz, R. P. Markus, and Z. P. Picarelli

*Department of Biochemistry and Pharmacology,
Escola Paulista de Medicina, C.P. 20372
01000 São Paulo, S.P., Brazil*

INTRODUCTION

During the experiments performed in Ca-free media on isolated rat vas deferens, it was observed that the contractions caused by epinephrine, acetylcholine (ACh), and serotonin fell to about 10% within 3 min, whereas the effects of barium chloride fell exponentially at a half-time of about 180 min. The decay of barium effects could be accelerated by giving successive 5-min doses of the agonist or by adding EDTA to the Ca-free medium. This chapter will analyze those results from the standpoint of a theoretical model for receptor mechanisms, based on Ca translocation, that has been developed by two of the authors (A. J. and R. P. M.). A detailed mathematical analysis of the model is beyond the scope of this volume.

The solution for the differential equations representing Ca transfer and drug-receptor interaction was obtained by numerical integration, by means of a programmed electronic calculator. The program is based on the assumptions currently accepted for occupation theory (Ariëns, 1964; Jurkiewicz, Jurkiewicz, Barros, and Valle, 1969; Jurkiewicz, Jurkiewicz, and Valle, 1971): in order to induce a contraction a drug (D) must interact with a specific receptor population (R); the resulting drug-receptor complex (DR) triggers a chain of events, leading to the contractile effect:

$$D + R \rightleftarrows DR \rightarrow a \rightarrow b \rightarrow \cdots \rightarrow l \rightarrow m \rightarrow n \rightarrow \text{effect.}$$

With this scheme in mind we hypothesized that one of the steps causes a Ca transfer into the cytoplasm from a membrane compartment that is in equilibrium with the external solution (Hurwitz and Suria, 1971; Chang and Triggle, 1972):

$$
\begin{array}{c}
\text{external Ca} \\
\downarrow \qquad \uparrow \\
\cdots \rightarrow l \rightarrow \text{bound Ca} \rightarrow \text{cytoplasm Ca} \rightarrow \text{effect} \\
\downarrow \\
\text{removed Ca}
\end{array}
$$

It was assumed that free Ca in cytoplasm is removed at a constant rate (equivalent to the "sink" mechanism of Hurwitz and Suria, 1971), but no assumptions were made about the nature of this Ca-removal system. It was also assumed that the effect is directly proportional to the concentration of free Ca in cytoplasm.

THEORY

Ca Pool

Studies performed with labeled Ca indicate that the ion is distributed in various kinetic compartments in smooth muscle (Lüllmann, 1970). In the present model only two Ca compartments are hypothesized: a superficial compartment, equivalent in the simplest case to the bath solution, and a sequestered compartment. Kinetics with Ca ligands is supposed to follow a Michaelis-Menten process, and therefore binding of external Ca (Ca_e) depends on the presence of free binding sites (B_f):

$$\frac{d(Ca_s)}{dt} = k_1 \, (Ca_e) \cdot (B_f) \tag{1}$$

and dissociation depends on the presence of bound calcium (Ca_s):

$$-\frac{d(Ca_s)}{dt} = k_2 \, (Ca_s) \tag{2}$$

where k_1 and k_2 are rate constants. Ca is considered to be loosely or tightly bound according to the values of these constants. Although several values of k_1 and k_2 were analyzed in order to conform to the half-times obtained by labeled Ca kinetics (Lüllmann, 1970), the figures presented here always refer to the same values for these constants. Unless otherwise stated, the results simulate steady-state conditions in Ca pools at the time a dose of agonist starts to interact with receptors.

Kinetics of Drug-Receptor Interaction

For the results reported here, the biophase barrier (Furchgott, 1964) and uptake mechanisms for catecholamines (Furchgott, 1972; Furchgott, Jurkiewicz, and Jurkiewicz, 1973) were supposed to be absent. For a given dose of agonist (D), the concentration of drug-receptor complex (DR) will attain a steady value according to association (k_3) and dissociation (k_4) constants (Waud, 1968):

$$\frac{d(DR)}{dt} = k_3 \cdot (D) \cdot (R) \tag{3}$$

$$-\frac{d(DR)}{dt} = k_4 \, (DR) \tag{4}$$

The capability of this complex to trigger the next event (a) will depend on a proportionality constant (k_5) similar to the intrinsic activity (Furchgott, 1964):

$$a = k_5(DR) \tag{5}$$

Ca Release

The event l is assumed to be directly proportional to step a and is supposed to cause the release of a quantum of bound Ca at each time-unit increment. This Ca enters the cytoplasm (Ca_c), and the effect is supposed to be directly proportional to its concentration:

$$\frac{d(Ca_c)}{dt} = l \cdot k_6 \, (Ca_s) \tag{6}$$

$$\text{Effect} = k_7 \, (Ca_c), \tag{7}$$

where k_6 and k_7 are proportionality constants. According to this model, an active interaction would therefore increase the rate of dissociation from Ca-binding sites. Although this is conceptually different from the assumption advanced by Hurwitz and Suria (1971) that an activation of Ca-transport sites occurs, it is equivalent to that hypothesis from an operational standpoint.

The removal mechanism is assumed to be proportional to Ca concentration in cytoplasm:

$$-\frac{d(Ca_c)}{dt} = k_8 \, (Ca_c) \tag{8}$$

where k_8 is a rate constant.

RESULTS

The main point of the present model is that the sequestered compartment is considered sufficiently small for its concentration to be significantly decreased by a prolonged drug-receptor interaction. Figure 1 shows theoretical time-response curves in which three interrelated events take place: formation of drug-receptor complex, release of Ca from a sequestered compartment, and entrance of Ca into the cytoplasm. The lower dose (*curve 5*) occupies about 10% of the receptors when equilibrium is reached, causing a proportional release of Ca from the sequestered compartment with a consequent increase in cytoplasm Ca. The highest dose, which occupies about 80% of receptors at equilibrium (*curve 1*), causing a larger release of Ca, greatly decreases bound Ca concentration, with a consequent decrease in the rate of Ca release (Eq. 6), inducing Ca concentration in cytoplasm to "fade" until a steady state is reached. Figure 2 shows theoretical time-

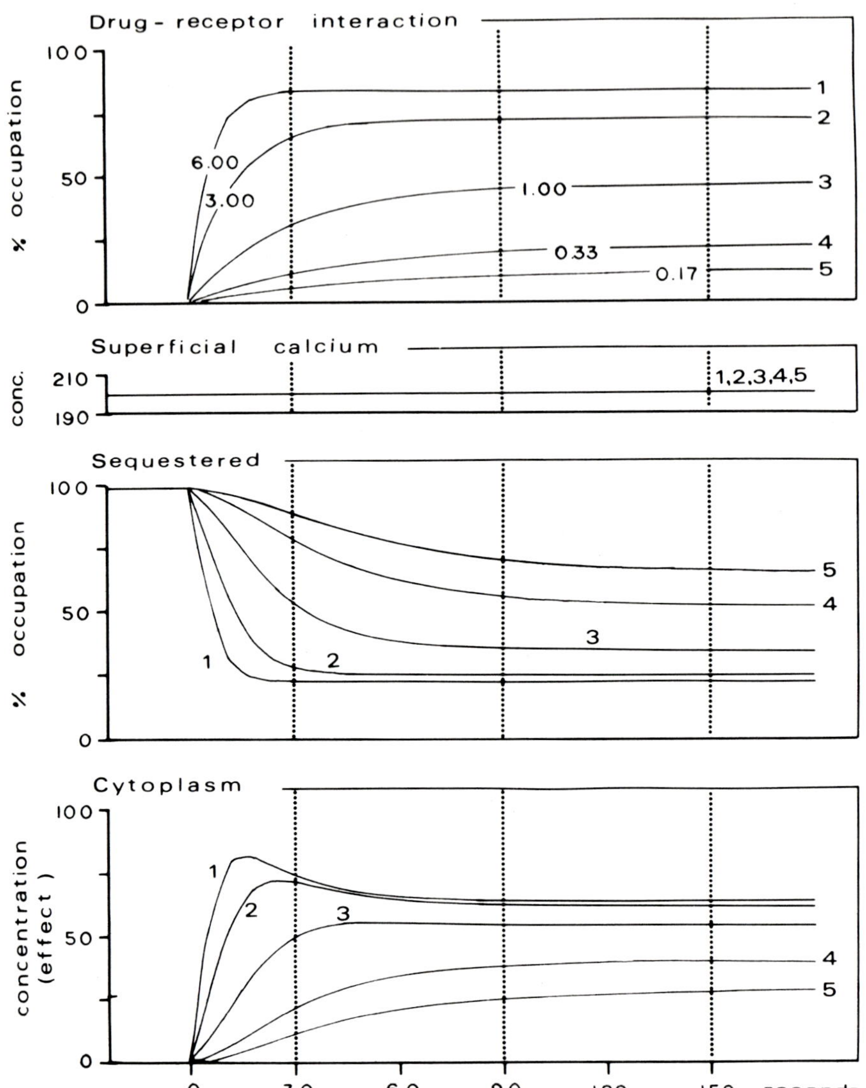

FIG. 1. Simulated time-response curves obtained by integration of the equations given in the text. The results expected from the use of 5 different doses were superimposed and the curves numbered correspondingly. *Top graph:* Doses are expressed as units of dissociation constant (D/K_A); thus curve 3 represents the effect of a dose occupying 50% of receptors after equilibrium of drug-receptor interaction. Simultaneously with drug-receptor interaction, other alterations take place in Ca concentrations, according to the corresponding numbers on the curves. Concentration in sequestered compartment is given as percentage of occupied binding sites. Superficial (extracellular) compartment is assumed to be large enough for its concentration to be constant (*middle graphs*).

response curves for six doses each of six hypothetical drugs obtained by varying the capability of the drug-receptor complex to generate step a (Eq. 5), namely by giving six different values for constant k_5. The effects decrease proportionally to k_5, a result equivalent to a variation in intrinsic activity. Note that drugs inducing low maximum effect are not expected to induce

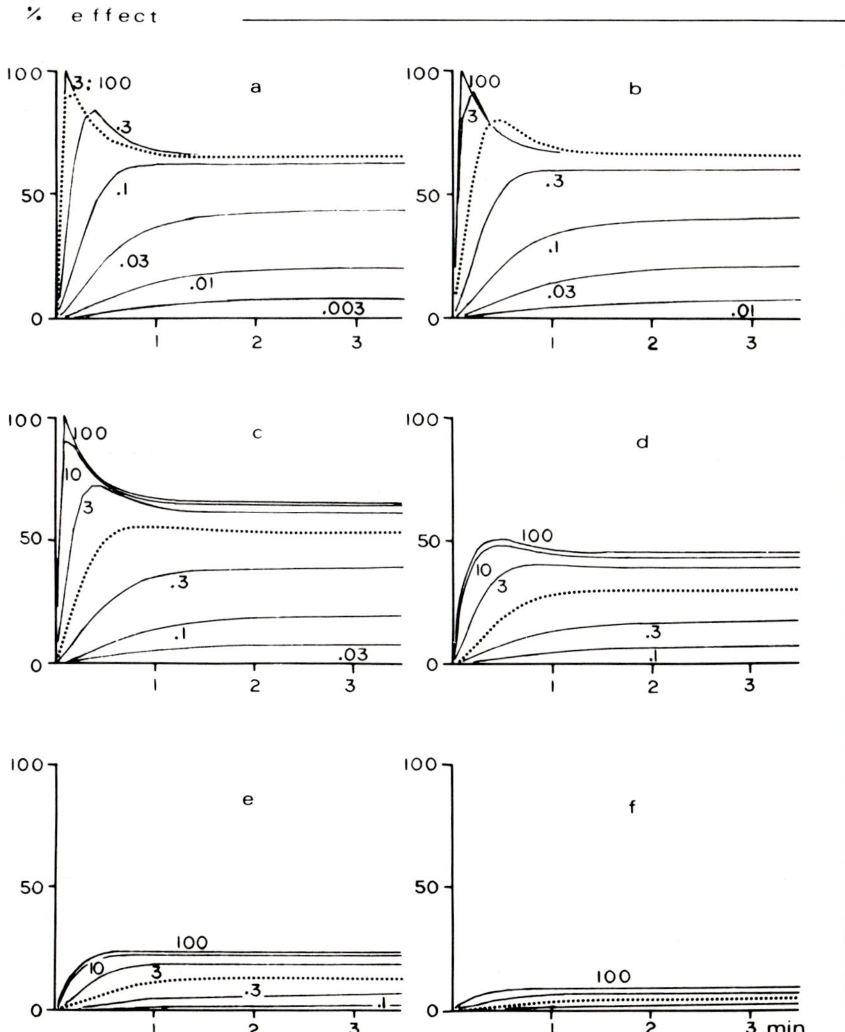

FIG. 2. Simulated time-response curves showing the influence of the ability of drug-receptor complex to release Ca. The value of constant k_5 varied as follows: a, 10.0; b, 3.0; c, 1.0; d, 0.3; e, 0.1; f, 0.03. Dotted time-response curves represent the effects of doses occupying 50% of receptors ($K_A = 1$). Doses are expressed in units of dissociation constants; thus, the highest dose (100 K_A) occupies almost all the receptors.

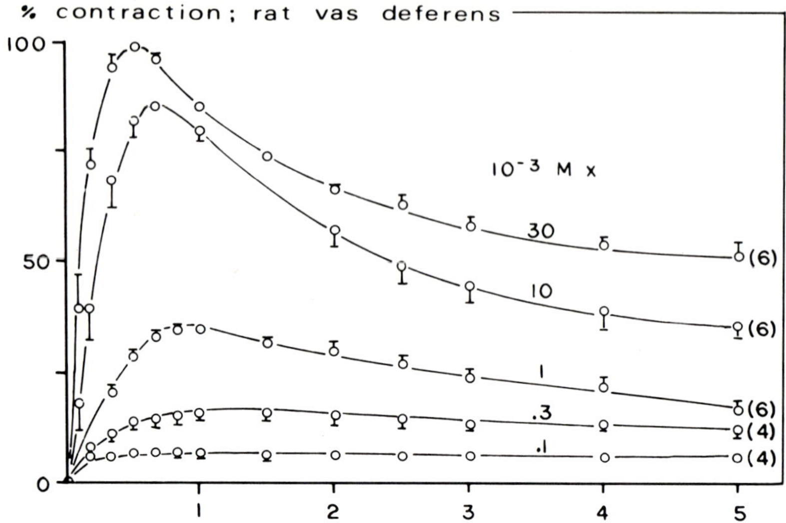

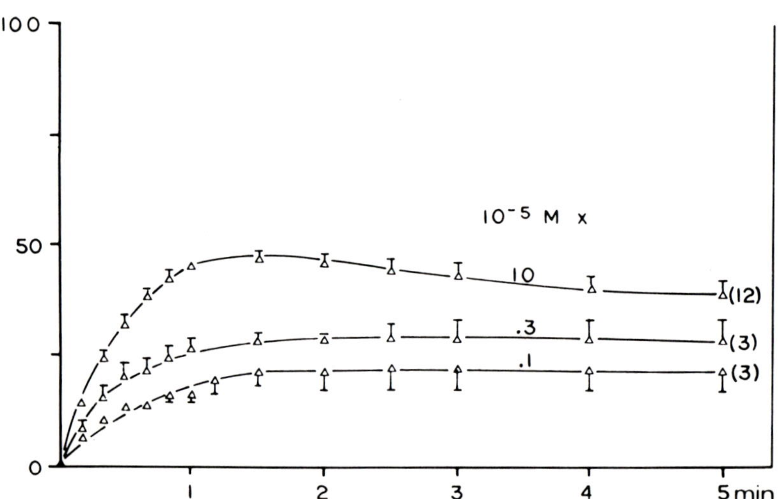

FIG. 3. Mean time-response curves for barium chloride (*top*) and epinephrine (*bottom*) in isolated vas deferens. Fading and the time necessary to attain peak response depend on the dose employed. Number of experiments in parentheses. Vertical lines represent SE.

fading, even by saturating the receptors. Although differences between the maximum effects of barium chloride and epinephrine cannot be ascribed to intrinsic activity (Jurkiewicz et al., 1969), the time-response curves for these agonists are in good agreement with the predicted curves (Fig. 2). This model also considers that when a drug is slowly added to the organ bath—for instance, by using a cumulative technique—the effect should be lower than that obtained when the agonist is added at once. As a matter of fact, this difference can be observed for epinephrine (Table 1) and barium chloride (Markus, Jurkiewicz, and Valle, 1971), as expected from theory.

TABLE 1. *Contractions induced by single or cumulative doses of epinephrine in rat vas deferens*

Dose (M)	Isotonic contractions (mm ± SEM)		No. of experiments
	Single doses	Cumulative doses	
3.10^{-7}	9.85 ± 2.54	9.74 ± 2.57	10
10^{-6}	25.48 ± 3.10	25.73 ± 3.17	10
3.10^{-6}	33.04 ± 2.35	32.90 ± 2.90	13
10^{-5}	44.92 ± 2.69	40.58 ± 2.72	13
3.10^{-5}	55.15 ± 3.89	41.88 ± 2.77[a]	10
10^{-4}	62.76 ± 4.61	46.20 ± 2.79[b]	10

[a] $p < 0.02$
[b] $p < 0.01$

Interesting results can also be obtained by simulating the increase or decrease of Ca concentration in the external medium. As a consequence, the degree of saturation of Ca-binding sites is expected to vary proportionally, causing a change in the shape of the time-response curves. One of the factors expected to change is the degree of fading. This prediction was confirmed by increasing Ca concentration from 1.8 mM to 36 mM. At the larger

TABLE 2. *Fading ratio for barium chloride (30 mM) at different external Ca concentrations*

Ca concn. in organ bath (mM)	No. of experiments	Fading ratio (FR)[a] (mean ± SEM)
1.8	69	0.55 ± 0.01
5.4	10	0.51 ± 0.02
7.2	24	0.48 ± 0.03
9.0	15	0.39 ± 0.02
18.0	10	0.23 ± 0.04
36.0	10	0.03 ± 0.03

[a] $FR = 1 - \dfrac{\text{contraction after 5 min}}{\text{peak contraction}}$

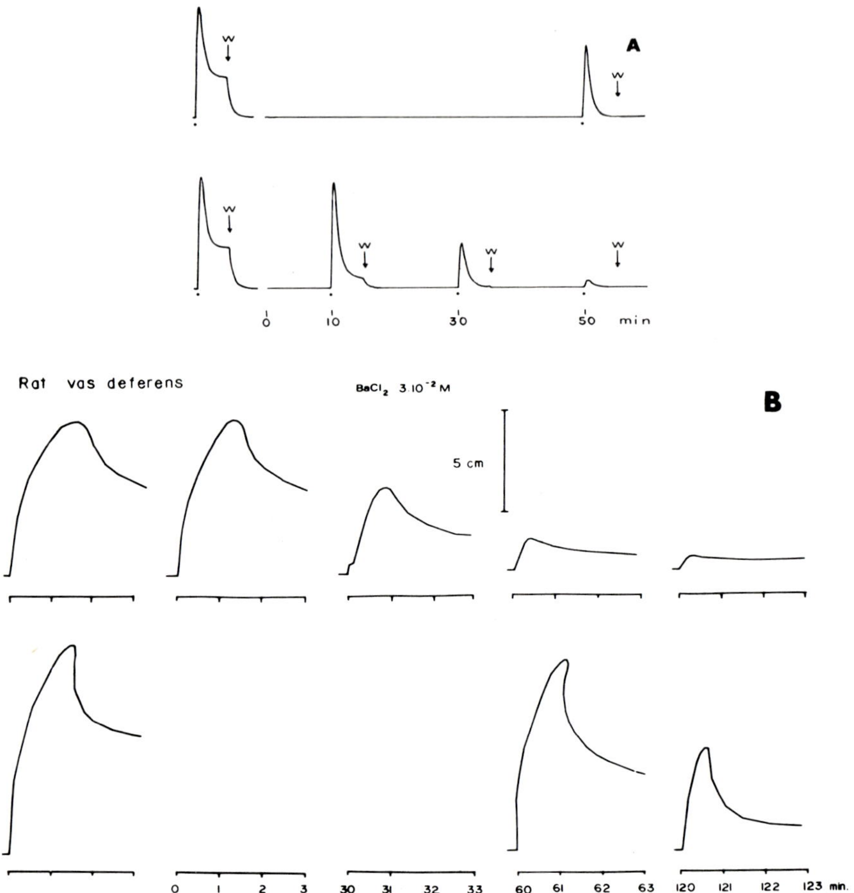

FIG. 4. (A) Simulated time-response curves for the effect of an agonist before and after Ca withdrawal from external medium (at zero min). Time units were adjusted in order to correspond approximately to the experimental results. In top tracing a drug-receptor interaction is supposed to take place 50 min after Ca withdrawal. Bottom tracing shows the effects expected when interactions take place at 20-min intervals. At W, drug withdrawal was simulated. (B) One out of six similar paired experiments in rat vas deferens in which barium chloride was tested after Ca withdrawal (0 min), either at 30-min intervals (*top*) or after a 60-min delay. The results were similar to the theoretical curves shown above.

concentration, practically no fading can be induced using barium chloride (Table 2). On the other hand, fading should be expected to increase following Ca withdrawal from the external solution (Fig. 3A). According to these theoretical curves, one should also expect the "peak" effect to be reduced according to the time of Ca omission and the number of doses of agonist. Figure 3B shows that although these last possibilities could be verified using

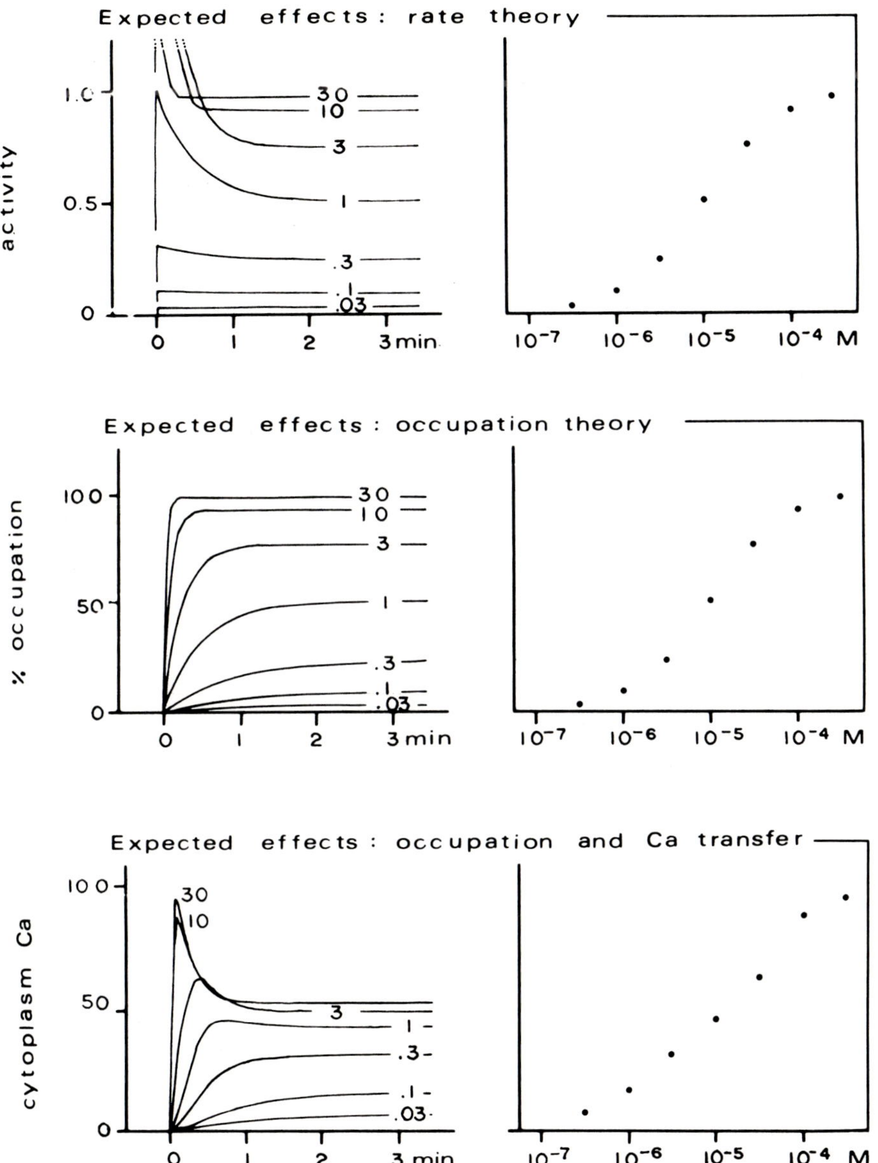

FIG. 5. *Graphs at left:* Time-response curves expected according to three different theoretical models for drug action. *Graphs at right:* Points in plots represent the contraction values at equilibrium (*top*), at peak responses (*bottom*), and when peak and equilibrium responses are the same (*middle*). Doses are expressed as k_A units. When peak and equilibrium responses are the same slope of the dose-response curve is not constant but fluctuates according to conditions of the Ca pool.

barium chloride, no decrease in fading was attained. However, full agreement with the theoretical curves was obtained with epinephrine in rabbit ear artery (Steinsland, Furchgott, and Kirpekar, 1973).

DISCUSSION

We believe that the main outcome of the model presented here is that Ca concentration in tissue pools is considered to vary according to dose and time of incubation of agonist, causing a change from the effects expected exclusively from occupation theory. As a consequence, the model accounts for most of the results obtained in rat vas deferens and explains "fading," which was not accounted for by occupation theory. Furthermore, other effects, such as specific and unspecific desensitization (Paton, 1961; Jurkiewicz et al., 1971) and receptor reserve (Ariëns, 1964), may be also analyzed.

It might be argued that fading could just as well be explained by rate theory. To analyze that possibility, theoretical time-response curves and dose-response curves expected according to rate theory, occupation theory, and the present model are shown in Fig. 5. Two effects must be considered in these curves: the "peak" and the "equilibrium" contractions, although both are coincident in the case of occupation theory. A sigmoid dose-response curve is obtained both for the Ca translocation model and for occupation theory. A sigmoid shape is also expected from rate theory when effects are measured at equilibrium. However, the peak effects are expected to increase indefinitely according to the dose. This is not the case for the results obtained in rat vas deferens, as shown in Fig. 3.

Notwithstanding the results obtained here, a quantitative analysis of the responses of rat vas deferens may still be necessary. For instance, it is difficult to explain why the effects of epinephrine fall almost immediately after Ca withdrawal—indicating the involvement of a loosely bound compartment—but show a variable rate of recovery after Ca reintroduction, which is proportional to the period of Ca omission. Another point concerns the differences between effects of cumulative and single doses; when epinephrine was used, these differences could be eliminated by increasing Ca concentration in organ bath, but similar results were not accomplished for barium chloride. On the other hand the effects of cumulative and single doses of barium were not significantly different in a potassium-rich solution. It is clear that additional work will be required for a better understanding of these last effects.

ACKNOWLEDGMENTS

This investigation was supported in part by the São Paulo State Research Foundation (FAPESP) and the Brazilian National Research Council (CNPq).

REFERENCES

Ariëns, E. J., editor (1964): *Molecular Pharmacology.* Academic Press, New York.

Chang, K. J., and Triggle, D. J. (1972): Membranal Ca^{++} translocation and cholinergic receptor activation. In: *The Role of Membranes in Metabolic Regulation,* pp. 59–110. Academic Press, New York.

Furchgott, R. F. (1964): Receptor mechanisms. *Annu. Rev. Pharmacol.* 4:21–50.

Furchgott, R. F. (1972): The classification of adrenoceptors (adrenergic receptors). An evaluation from the standpoint of Receptor Theory. In: *Catecholamines, Handb. Exp. Pharmacol.,* Vol. 33, edited by H. Blaschko & E. Muscholl, pp. 283–335. Springer-Verlag, Berlin and New York.

Furchgott, R. F., Jurkiewicz, N. H., and Jurkiewicz, A. (1973): Analysis of propranolol antagonism to isoproterenol in guinea-pig trachea before and after blockade of active uptake. *Fed. Proc.* 32:723 Abs.

Hurwitz, L., and Suria, A. (1971): The link between agonist action and response in smooth muscle. *Annu. Rev. Pharmacol.* 11:303–326.

Jurkiewicz, A., Jurkiewicz, N. H., Barros, G. S., and Valle, J. R. (1969): Relative responsiveness (ρ) of pharmacological receptor systems in the rat *vas deferens. Pharmacology* 2:89–99.

Jurkiewicz, N. H., Jurkiewicz, A., and Valle, J. R. (1971): Contractile reserve after maximum doses of full agonists on the rat vas deferens. *Pharmacology* 5:129–144.

Lüllmann, H. (1970): Calcium fluxes and calcium distribution in smooth muscle. In: *Smooth Muscle,* edited by E. Bülbring, A. F. Brading, A. W. Jones, and T. Tomita, pp. 151–165. Edward Arnold, London.

Markus, R. P., Jurkiewicz, A., and Valle, R. R. (1971): Comparação entre doses simples e cumulativas de alguns agonistas totais na musculatura lisa isolada. *Ciência Cult.* (São Paulo) 23:384.

Paton, W. D. M. (1961): A theory of drug action based on the rate of drug-receptor combination. *Proc. R. Soc. Lond. [Biol]* 154:21–69.

Steinsland, O. S., Furchgott, R. F., and Kirpekar, S. M. (1973): Biphasic vasoconstriction of the rabbit ear artery. *Circ. Res.* 32:49–58.

Waud, D. R. (1968): Pharmacological receptors. *Pharmacol. Rev.* 20:49–88.

Concepts of Membranes in Regulation and Excitation,
edited by M. Rocha e Silva and G. Suarez-Kurtz.
Raven Press, New York © 1975.

Phase-Plane Determination of Membrane Currents in Propagated Action Potentials: Possibilities and Difficulties

Antonio Paes de Carvalho

Institute of Biophysics, Center for Health Sciences, Federal University of Rio de Janeiro, Rio de Janeiro, Brazil

INTRODUCTION

The display of transmembrane-propagated action potentials in the phase plane (membrane voltage versus its first time derivative) is a simple technique that can yield much valuable information. It has been used by Wallon (1963), by Paes de Carvalho, Hoffman, and Langan (1966), and by Paes de Carvalho, Hoffman, and Paula Carvalho (1969) to study details of voltage-time course during the action potential. Jenerick (1963, 1964) showed that with the admission of some simplifying assumptions it was possible to obtain an estimate of membrane ionic current during the action potential by the use of the phase plane. This chapter extends Jenerick's original treatment and discusses the applicability of the method to skeletal and cardiac muscle.

PHASE-PLANE RECORDING OF PROPAGATED ACTION POTENTIALS

Action potentials can easily be recorded in the phase plane with the aid of a suitable differentiator and an X-Y oscilloscope channel, as shown in Fig. 1 (see also Paes de Carvalho et al., 1969). Recording of membrane voltage is effected through a regular 3M KCl-filled glass microelectrode (20 to 30 megohm tips) and a suitable capacity-compensated, high-input impedance electrometer. The membrane voltage thus obtained is fed to the horizontal axis of the X-Y scope and to the input of an electronic differentiator (an operational amplifier with appropriate input and feedback connections and frequency response). The frequency response of the whole system is adjusted for optimal reproduction of either sine waves or ramps, with good fidelity and a minimum of noise in the derivative record. Inappropriate compensation at the electrometer or at the differentiator causes distortion of the record and must be carefully avoided. A phase loop recorded during

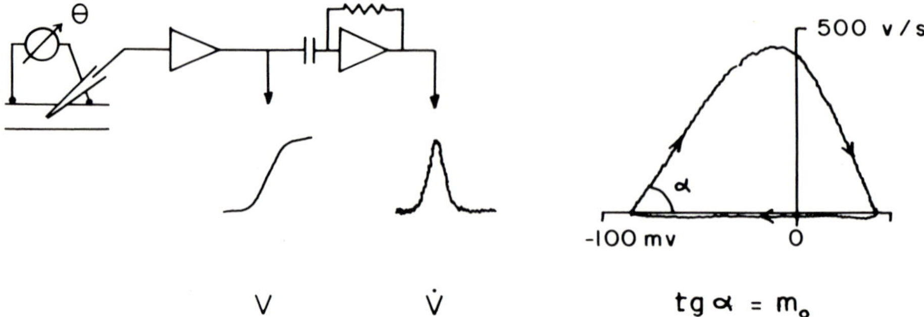

FIG. 1. Phase-plane recording of transmembrane propagated action potentials. *Left to right:* Microelectrode and surface electrode placements for measuring membrane voltage, *V,* and conduction velocity, *θ*; electrometer preamplifier and electronic differentiator to yield *V* and *V̇*; and phase loop recorded during the action potential. The record departs from resting potential (*extreme left of loop*) and courses clockwise. Note initial slope *m_o*, which equals the rate constant of the exponential foot of the action potential.

a propagated action potential in heart muscle is schematically represented in Fig. 1. Measurement of the slope of the phase record at different voltages is an important part of the method to be described. This slope is generally called *m* and the slope of the initial portion of the phase loop is called *m_o*. Note that the initial portion of the loop corresponds to the subthreshold region at the foot of the action potential.

MATHEMATICAL BASIS FOR COMPUTING MEMBRANE CURRENTS FROM PHASE LOOPS

Jenerick (1963, 1964) has shown that it is possible to integrate Hodgkin and Huxley's (1952) equation for the propagated action potential to obtain the instantaneous ionic current (I_i) at any given time:

$$I_i = C\dot{V}\left(\frac{m}{k} - 1\right) \tag{1}$$

In Eq. 1, the ionic current (in A/cm^2) appears as a function of membrane capacitance (C, F/cm^2), of dV/dt ($\dot{V}$, volts/sec) and m at any given instant, and of the constant k of the propagated action potential equation (equal to $2R_iC\theta^2/a$, where R_i is the internal resistivity in ohm·cm, θ the conduction velocity in cm/sec, and a is fiber radius in cm). Since membrane ionic conductance at the foot of the skeletal muscle action potential (i.e., at rest) is more than a hundred times smaller than the ionic conductance during activity, Jenerick assumed that m_o, the slope of the initial straight portion of the phase loop, was equal to k. This assumption implies that membrane current at the foot of the action potential is all capacitive and that membrane resistance (R_m in ohms·cm^2) is infinite. The assumption enables membrane ionic current to be computed from Eq. 1, provided C is known. Jenerick (1963, 1964) has applied the method to frog skeletal muscle fibers, where the geometry and constant propagation requirements seemed to hold. Paes de Carvalho et al. (1969) have attempted its use in cardiac muscle (rabbit atrial trabecula), based on the assumption that a cardiac muscle bundle will approach cable behavior when invaded by a flat wave front (i.e., all fibers in a given cross section are at the same voltage throughout the action potential).

We have extended Jenerick's mathematical treatment by avoiding the assumption that $k = m_o$. In this fashion, R_m will no longer be infinite and values can be obtained for membrane currents at subthreshold voltages. The main advantages should be a more correct estimate of threshold and the possibility of simultaneously computing the resting membrane time constant and space constant. On the other hand, this development poses the problem of the experimental determination of the constant k, the value of which is smaller than but quite close to that of m_o (differences may amount to less than 1% when active conductance is much higher than resting conductance). For the purpose of our studies, we have considered membrane ionic currents per unit membrane capacitance and not per unit membrane area. Equation 1 then converts to:

$$I = \frac{I_i}{C} = \dot{V}\left(\frac{m}{k} - 1\right) \tag{2}$$

The units of I are therefore A/F or a suitable multiple.

It is well known that the foot of the propagated action potential has a simple exponential time course (Tasaki and Hagiwara, 1957) and is represented on the phase plane by the rectilinear segment departing from the resting potential. The slope m_o is the rate constant of that exponential. During the inscription of the straight line, it can be said that the membrane ionic channels are at rest and exhibit a linear behavior in response to current injected by the advancing wavefront. From the exponential character of the foot, it can be said that at any instant the membrane potential, V, during the foot equals the product $m_o\dot{V}$. The time constant (τ) of the resting membrane can thus be computed:

$$\tau = R_m C = \frac{VC}{I_i} = \frac{V}{I} = \frac{1}{m_o}\left(\frac{m_o}{k} - 1\right) \tag{3}$$

This relationship can be used to determine the constant k experimentally. For this, it is sufficient to secure measurements of m_o and conduction velocity (θ) in a cell during exposure to two experimental situations, one being the "normal" (control) and another providing a small change in m_o

and θ *without altering the resting membrane time constant.* In this case, it can be said from Eq. 3:

$$m_o \left(\frac{m_o}{k} - 1 \right) = m_o' \left(\frac{m_o'}{k'} - 1 \right),$$

where m_o' and k' are the values obtained in the test solution.

Supposing that R_i, C, and a remain constant, one has:

$$k' = \frac{\theta'^2}{\theta^2} \, k = \frac{k}{Z},$$

where Z stands for $\dfrac{\theta^2}{\theta'^2}$; and:

$$k = \left(\frac{m_o^2 - Z m_o'^2}{m_o - m_o'} \right) \tag{4}$$

It should be noted here that the accuracy of the determination of k depends heavily on the accuracy of the determination of conduction velocities (a precision of one part in a thousand is necessary to keep errors of estimated time constant within 10%).

Once the time constant has been obtained, the space constant can be computed using the following relationship:

$$\lambda = \frac{\theta}{m_o} \sqrt{1 + m_o \tau} \tag{5}$$

The foregoing relationship can be easily derived from cable behavior, assuming an exponential time course with rate constant m_o for the membrane voltage at the foot of the action potential. It should be noted that the ratio θ/m_o is the space constant of the exponential foot of the advancing action potential (which is of course shorter than the true λ).

PHASE-PLANE ANALYSIS OF MEMBRANE CHARACTERISTICS IN THE RABBIT ATRIUM

The method outlined above was applied to rabbit atrial muscle in regions where propagation was thought to be simple and wave shape constant. Activity originated from a distant stimulating electrode and invaded the site of transmembrane recording, and conduction velocity was measured along a flat, broad front. Conduction velocity was measured in a surface bipolar record as in Fig. 1. Under these conditions, cable behavior was assumed for the tissue.

Figure 2 shows a record taken during an experiment where data from control Tyrode solution were compared with data obtained under a low

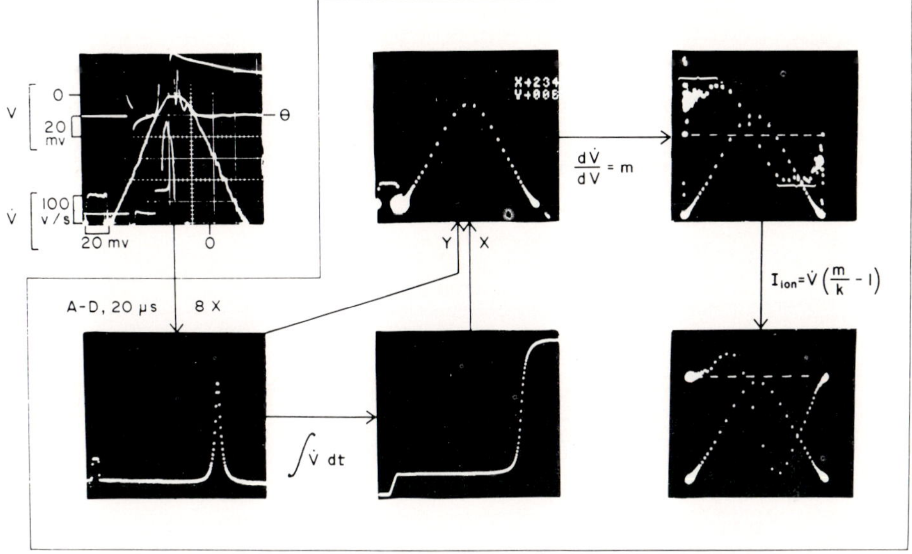

FIG. 2. *Top left record:* Oscilloscope records of surface electrogram (θ), membrane voltage-time course (*V*), and phase-plane display of a rabbit atrial cell. The electronically obtained $\dot{V}$ was A–D converted at a rate of 50 KH (sample timing locked with stimulus) and the average of 8 successive cycles taken. From this the phase loop was regenerated and, after entering the value of *k*, the *I,V* plot for membrane ionic current was obtained. (Obtained with program prepared by S. V. Almeida, 1972.)

Na concentration (86 mM choline chloride substitution). In both cases a strong acetylcholine (Ach) background was maintained (16 μg/ml) to swamp any possible effects of choline on the membrane constants. Measurements can be made visually from the oscilloscope records but are rather tedious if membrane currents throughout the action potential upstroke are to be computed. This inconvenience was circumvented by the on-line utilization of a small digital computer (Digital Equipment Corporation, PDP-12). Samples of either membrane voltage or its first-time derivative (analogically obtained) could be taken at 20-μsec intervals.

For Fig. 2, sampling of the first time derivative was used. Membrane voltage was obtained by integration with a constant chosen to yield a stable resting potential. The phase loop of the action potential can thus be reconstructed and the value of *m* determined for any value of *V* (m_o is the initial value for *m*). Measurements taken in the low Na solution can then be compared. The surface potentials are also sampled and compared in order to obtain values for conduction velocity in both solutions. The constant *k* can then be computed through Eq. 4 and entered together with $\dot{V}$ and *m* in Eq. 2 to obtain a plot of membrane current (per unit capacitance) versus voltage during the upstroke of the action potential. Time and space constants for the resting membrane can also be obtained from Eqs. 3 and 5. Values obtained in this fashion from three different cells in which a stable microelectrode impalement was kept throughout the experiment are shown in Table 1. As can be seen, there is considerable spread in the data obtained from these three cells, but only in the second cell are the values out of range with data obtained by other methods (Woodbury and Crill, 1961; Bonke, 1973).

TABLE 1. Passive constants of rabbit atrial muscle obtained by the phase-plane method

Cell	Time constant (msec)	Space constant (mm)
1	1.21	0.43
2	0.31	0.18
3	1.82	0.38

It should be pointed out that the space and time constants shown in Table 1 can be expected to fall somewhat below other values available in the literature, due to the high Ach background used in our experiments (Trautwein, 1963). However, this effect may be less important than we think, owing to the fact that action potentials from isolated rabbit atria show an appreciable degree of recovery from Ach effects if exposure is prolonged beyond 5 min. Recovery is almost complete after 20 or 30 min of exposure (Paula Carvalho, 1972).

CONTRIBUTION OF Na IONS TO MEMBRANE CURRENT DURING THE INITIAL FAST COMPONENT OF THE CARDIAC ACTION POTENTIAL

It has been shown that the cardiac action potential is composed of two independent and separable voltage components (Paes de Carvalho et al., 1966, 1969). The first, called the "fast component," contributes the initial fast depolarization and is short lived. The second, the "slow component," has a slow rate of rise that is usually masked by the fast component but lasts considerably longer, yielding the familiar plateau of cardiac action potential. Today it is commonplace to say that these two components are the operation of different ionic channels (Trautwein, 1973)—namely, a fast Na channel with a Hodgkin-Huxley type of voltage and time dependence, and a slow channel for Na and Ca ions that shows a different behavior and a peculiar drug sensitivity.

Analysis of the inward current observed during an action potential (Fig. 3) was carried out by the phase-plane method, in an attempt to confirm the

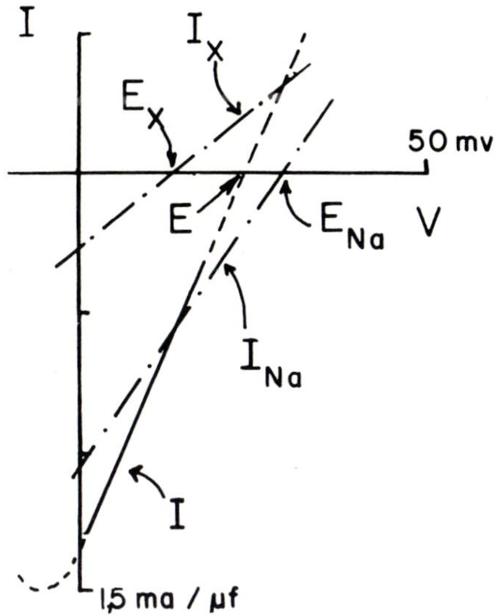

FIG. 3. Contribution of Na⁺ current (I_{Na}) and other ionic currents (I_x) to total current (I) during the last 40 mV of the upstroke in rabbit atrium. E_{Na}, Na equilibrium potential. E_x, combined equilibrium potential for all other ions. See text.

nature of the ion involved. This is easily done by observing the effects of two modified solutions on membrane current. One of them is the low Na (86 mм) solution already mentioned. The other is a solution containing a concentration of tetrodotoxin (TTX) (1 µg/ml) that can be considered low for mammalian cardiac muscle. The analysis is based on the fact that membrane ionic current seems to depend linearly on membrane voltage in the last 30 mV of the action potential upstroke (Fig. 2). That can be construed to mean that membrane chord conductance per unit capacitance (G) is constant in this voltage range, throughout which the membrane tends towards a fixed equilibrium potential (E). The membrane conductance at this time is the sum of the Na⁺ conductance (G_{Na}) and all other conductances combined (G_x). G_{Na} can be assumed constant and maximally activated in this voltage range if it is to follow a Hodgkin-Huxley behavior. As a consequence, G_x must be assumed also constant.

Therefore, during the rectilinear portion of the I, V plot at the end of the upstroke, one can write:

$$I = G(V - E);$$

and

$$I = I_{Na} + I_x = G_{Na}(V - E_{Na}) + G_x(V - E_x).$$

Then when the low Na solution is applied, E_{Na} changes to a new value. Even though the absolute value of E_{Na} was not known beforehand, the change (ΔE_{Na}) resulting from Na reduction can be calculated from the

Nernst equation, assuming a constant internal Na. However, it is not possible to rule out *a priori* a change in G_{Na}.

Taking these considerations into account, we may write, for any given voltage, V, within the range studied:

$$\Delta I = \Delta G(V - E_{Na}) + G'_{Na} \cdot \Delta E_{Na} \tag{6}$$

where ΔI, ΔG and ΔE_{Na} represent the change in these parameters resulting from the experimental maneuver. The change in total current and conductance is measured on the I, V plot and the change in E_{Na} can be calculated. Equation 6 is therefore left with two unknowns, namely E_{Na} (control Na equilibrium potential) and G'_{Na} (active Na conductance in the low Na solution). The control value for E_{Na} can be obtained from data with the TTX solution, which causes a change in G_{Na} without altering E_{Na} or G_X. Comparing this data with that in normal Tyrode, one can write for a given V:

$$E_{Na} = V - \left(\frac{\Delta I}{\Delta G}\right) \tag{7}$$

The value of E_{Na} can now be entered in Eq. 6, which can thus be solved for G'_{Na}. The I, V plot for the activated Na channel can then be obtained. The I, V plot for the combination of other ions is obtained as the difference between this and the total current.

One cell completely studied in this fashion showed that there is no appreciable change in active G_{Na} with the low Na solution used. Data calculated from the experiment were as follows: Resting characteristics, $\lambda = 0.23$ mm and $\tau = 0.6$ msec. Ionic characteristics during maximal Na^+ channel activation were: $G = 22.8$ mmho/μF; $G_{Na} = 15.8$ mmho/μF; $G_X = 7.0$ mmho/μF; $E_{Na} = 27.5$ mV; $E_X = 13$ mV; $[Na^+]_i = 33$ mM.

Time and space constants for the cell are probably underestimated. This may have resulted in overestimation of total conductance (G) during the last 40 mV of the action potential upstroke, with proportionate increases in the estimates for G_{Na} and G_X during the maximal activation of the Na channel. But equilibrium potentials for Na and X should not be affected by those inaccuracies. The values observed for E_{Na} and internal Na^+ concentration are in keeping with measurements performed by Goodford and Vaughan Williams (1962) on the same preparation. It is surprising to find that E_X, the combined equilibrium potential for all other ions, has a positive value in this range. That may indicate considerable activation of a Ca^{++} current early in the action potential. Of course, the results of the analysis would be subject to considerable restriction if the X channel did not show constant conductance within the voltage range studied. However, this possibility would be in keeping with a linear I, V plot for the X channel only if the contributing conductances were to change with voltage in such a fashion as to yield a linear resultant.

CRITICISM OF THE APPLICATION OF THE PHASE-PLANE METHOD TO HEART AND SKELETAL MUSCLE

The type of analysis carried out above requires (a) perfect cable behavior in the structure studied; (b) a simple resistance-capacitance parallel arrangement for the membrane model, the capacitance remaining constant throughout the measurements; and (c) constant conduction velocity and wave shape throughout a segment of tissue several space constants long at each side of the measuring microelectrode. The first two requirements have been assumed valid for the invasion of the atrial muscle cell by a fast wave shape such as the upstroke of the action potential. The last requirement is not easy to establish for the case of constant wave shape but can be admitted to be observed in either skeletal muscle or a length of "uniform" atrial muscle. Given these conditions, the Hodgkin-Huxley formulation of the propagated action potential yields in the phase plane a straight line corresponding to the foot of the action potential (Saldeña, Garcia, and Paes de Carvalho, 1973). The slope of this straight line is *the largest positive slope* of the phase plane during the upstroke of the action potential. Threshold is defined as the voltage at which the decreasing m becomes equal to the constant k.

A large number of phase-plane recordings have been obtained from cardiac and skeletal muscle cells in our laboratory during the past few years. It is now apparent that a large proportion of cells (more than 70%) yield phase loops in which the initial linear segment breaks after some 20 to 30 mV into *a second segment of steeper slope* (Garcia, Saldeña, and Paes de Carvalho, 1973). This is shown for cardiac muscle in Fig. 4 and for skeletal muscle in Fig. 5. The same pattern can be detected in the record and computer displays shown in Fig. 2 and in Jenerick's (1963) data. We do not yet understand the meaning of this finding, beyond the fact that it renders our use of the phase plane open to criticism. On the other hand, we feel that the observation is important in itself as it shows that propagation in cardiac and skeletal muscle cells may not follow the Hodgkin-Huxley

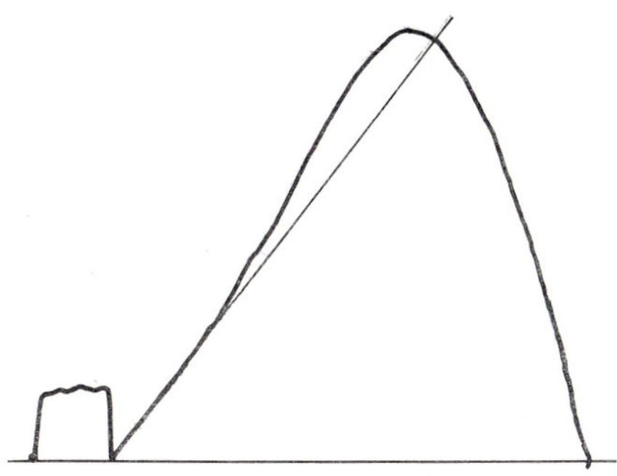

FIG. 4. Phase record from the upstroke of action potential in a rabbit Purkinje fiber. Note that the initial slope is not the maximum positive slope. Calibration at left end of record: 20 mV (horizontal), 100 v/sec (vertical). (Obtained by E. C. Garcia, 1974.)

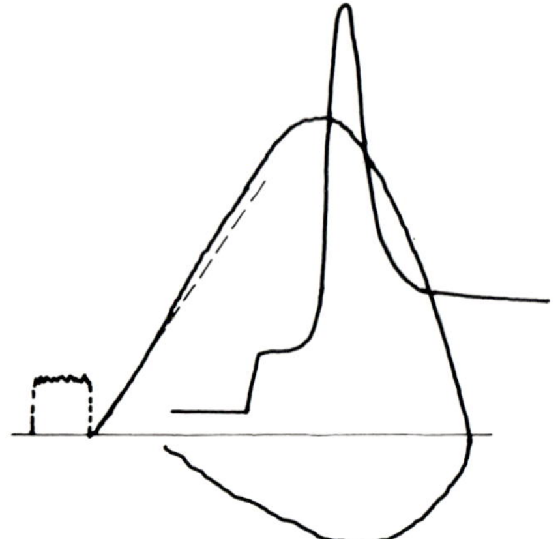

FIG. 5. Phase record from the upstroke of the action potential in frog skeletal muscle. Note that the initial slope is not the maximum positive slope. Calibration at left end of record: 20 mV (horizontal), 100 v/sec (vertical). (T. A. Saldeña, 1974.)

equation for the propagated action potential (which was, of course, developed for squid axon), even when only the foot and upstroke of the action potential are considered.

ACKNOWLEDGMENTS

This work was supported by grants from the Brazilian National Research Council (CNPq), the Council for Graduate Education of the Federal University of Rio de Janeiro, and the Brazilian National Bank for Economic Development (FUNTEC 241).

The author is grateful to Dr. Carlos Eduardo Rocha Miranda for having introduced him to the PDP-12 computer; and to Dr. Sergio Verjovski de Almeida for having developed the programs used in the phase-plane analysis of the cardiac action potential.

REFERENCES

Bonke, F. I. M. (1973): Passive electrical properties of atrial fibers of the rabbit heart. *Pfluegers Arch.* 339:1–15.

Garcia, E. C., Saldeña, T. A., and Paes de Carvalho, A. (1973): Discrepância entre teoria e realidade na análise do potencial de ação cardíaco pelo plano de fase. *An. Acad. Bras. Cienc.* 45(3/4):670–671.

Goodford, P. S., and Vaugham Williams, E. M. (1962): Intracellular sodium and potassium concentrations of rabbit atria, in relation to the action of quinidine. *J. Physiol. (Lond.)* 160:483–493.

Hodgkin, A. L., and Huxley, A. F. (1952): A quantitative description of membrane current and its application to conduction and excitation in nerve. *J. Physiol. (Lond.)* 117:500–544.

Jenerick, H. (1963): Phase-plane trajectories of the muscle spike potential. *Biophys. J.* 3:363–377.

Jenerick, H. (1964): An analysis of the striated muscle fiber action current. *Biophys. J.* 4:77–91.

Paes de Carvalho, A., Hoffman, B. F., and Langan, W. B. (1966): Two components of the cardiac action potential. *Nature* 211:938–940.

Paes de Carvalho, A., Hoffman, B. F., and Paula Carvalho, M. (1969): Two components of the cardiac action potential. I. Voltage-time course and the effect of acetylcholine on atrial and nodal cells of the rabbit heart. *J. Gen. Physiol.* 54:607–635.

Paula Carvalho, M. (1972): Influencia da Frequencia de Estimulação sobre a Forma do Potencial de Ação Cardíaco. *Thesis,* Instituto de Biofísica, Universidade Federal do Rio de Janeiro, Brasil.

Saldeña, T. A., Garcia, E. C., and Paes de Carvalho, A. (1973): Plano de fase do modelo de Hodgkin e Huxley para o potencial de ação do axônio gigante da lula. *An. Acad. Bras. Cienc.* 45(3/4):670–671.

Tasaki, I., and Hagiwara, S. (1957): Demonstration of two stable potential states in the squid giant axon under tetraethylammonium. *J. Gen. Physiol.* 40:859–885.

Trautwein, W. (1963): Generation and conduction of impulses in the heart as affected by drugs. *Pharmacol. Rev.* 15:277–330.

Trautwein, W. (1973): Membrane currents in cardiac muscle fibers. *Physiol. Rev.* 53:793–835.

Wallon, G. (1963): Etude cinétique du potentiel d'action du myocarde ventriculaire. I. Principe et technique. *C. R. Soc. Biol. (Paris)* 157:2148–2151.

Woodbury, W. J., and Crill, W. E. (1961): On the problem of impulse conduction in the atrium. In: *Nervous Inhibition,* edited by E. Florey, pp. 124–135. Pergamon Press, New York.

Concepts of Membranes in Regulation and Excitation,
edited by M. Rocha e Silva and G. Suarez-Kurtz.
Raven Press, New York © 1975.

Active Sodium Transport and Oxygen Coupling in Amphibian Skin

F. Lacaz Vieira and G. Danisi

Department of Physiology and Pharmacology, Institute of Biomedical Sciences, University of São Paulo, 01000 São Paulo, S.P., Brazil

INTRODUCTION

The isolated amphibian skin has been used as a model for the study of the genesis of bioelectrical potentials and ion transport since the end of the 19th century (see reviews by Dean and Gatty, 1937; Linderholm, 1952; and Huf, 1955). The work of Ussing, his collaborators, and followers brought the preparation of amphibian skin to an important place in the literature dealing with ion transport and bioelectrical phenomena in epithelial membranes (Ussing, 1953, 1954, 1960; Ussing, Erlij, and Lassen, 1974).

The isolated amphibian skin can be treated on a thermodynamic basis as a system that generates an electrical potential difference and accomplishes ion transport at the expense of free energy coupled to those processes. There have been several studies of the coupling of ion transport and potential difference in amphibian skin with the oxidative metabolism (Conway, 1951, 1953, 1955; Leaf and Renshaw, 1957; Zerahn, 1956, 1958; Vieira, Caplan, and Essig, 1972a,b; Danisi and Lacaz Vieira, 1974).

NONEQUILIBRIUM THERMODYNAMIC FORMALISM

Essig and Caplan (1968) have developed a theoretical nonequilibrium thermodynamic description of active transport of a cation and its coupling to a metabolic driving reaction. This formalism describes the behavior of the transport system in terms of phenomenological equations and assumes a linear dependence (in the experimental range of validity of these equations) between the rate of active sodium transport and the flux of the metabolic driving reaction with the Na electrochemical potential difference across the skin. The necessary equations derived from the work of Essig and Caplan (1968) will be presented in terms of conductance coefficients.

$$J_{\mathrm{Na}}^{a} = L_{\mathrm{Na}} X_{\mathrm{Na}} + L_{\mathrm{Na}r} A \tag{1}$$

$$J_{r}^{*} = L_{r\mathrm{Na}} X_{\mathrm{Na}} + L_{r} A \tag{2}$$

J_{Na}^{a} represents net active sodium flux; J_{r}^{*} is the suprabasal rate of oxygen consumption—equal to total rate of oxygen consumption, (J_r) minus the rate of oxygen consumption obtained in the absence of short-circuit current, $(J_r)_{I_0=0}$; I_o is the short-circuit current; X_{Na} is the negative electrochemical potential difference of sodium across the skin; A is the affinity of the metabolic driving reaction; L is a phenomenological conductance coefficient. The validity of Onsager's relationship has been assumed ($L_{\mathrm{Na}r} = L_{r\mathrm{Na}}$). X_{Na} can be expressed explicitly as:

$$X_{\mathrm{Na}} = RT \ [\ln(\mathrm{Na})_e - \ln(\mathrm{Na})_i] - F\Delta\psi \tag{3}$$

where: $(\mathrm{Na})_e$ and $(\mathrm{Na})_i$ are the sodium concentrations in the external and internal bathing solutions; $\Delta\psi$ is the electrical potential difference across the skin, equal to $\psi_i - \psi_e$; and R, T, and F have their conventional meanings.

Alterations of A or of X_{Na} should induce variations in the fluxes J_{Na}^{a} and J_{r}^{*}, which are expected to be linear functions of the forces (A or X_{Na}) in the range of experimental validity of these equations. In this range, both fluxes are expected to be linear functions of the electrical potential difference and logarithmic functions of the sodium concentration in the bathing solutions.

In the preparation of amphibian skin it is possible to control experimentally X_{Na} by given values of $(\mathrm{Na})_e$, $(\mathrm{Na})_i$, and $\Delta\psi$. On the other hand, the affinity, A, cannot be controlled experimentally, but it can be evaluated in some circumstances (Essig and Caplan, 1968; Vieira et al., 1972a; Saito, Essig, and Caplan, 1973).

The phenomenological coefficients can be evaluated according to the following equations, which are adequate to the specified experimental conditions:

$$L_{\mathrm{Na}} = \frac{1}{RT} \left[\frac{\Delta J_{\mathrm{Na}}^{a}}{\Delta \ln(\mathrm{Na})_e} \right] (\mathrm{Na})_i, \ A \tag{4}$$

$$L_{r\mathrm{Na}} = \frac{1}{RT} \left[\frac{\Delta J_{ro}^{*}}{\Delta \ln(\mathrm{Na})_e} \right] (\mathrm{Na})_i, \ A \tag{5}$$

$$L_{\mathrm{Na}r} = L_{r\mathrm{Na}} \ \text{(by Onsager's relationship) and} \tag{6}$$

$$L_r = L_{\mathrm{Na}r} \left[\frac{1}{\Delta J_{\mathrm{Na}}^{a}/\Delta J_{ro}^{*}} \right] (\mathrm{Na})_e = (\mathrm{Na})_i \tag{7}$$

where J_{ro}^{*} is the suprabasal rate of oxygen consumption measured in the short-circuited state; $\Delta J_{\mathrm{Na}}^{a}/\Delta J_{ro}^{*}$ is the stoichiometric ratio observed during spontaneous decline of sodium transport in the condition of zero electrochemical potential difference across the skin, due to identical sodium concentrations on both sides and $\Delta\psi$ clamped to zero, equal to $L_{\mathrm{Na}r}/L_r$. When $\Delta\psi = 0$ and sodium concentration is reduced in the external solution, the stoichiometric ratio is equal to $L_{\mathrm{Na}}/L_{r\mathrm{Na}}$.

METHODOLOGY

Abdominal amphibian skins were mounted between Lucite hemichambers and bathed on both sides by solutions of different compositions. A voltage clamp was used to set the potential difference across the skin, and the clamping current was recorded continuously. The partial pressure of oxygen in each bathing solution was continuously monitored by means of Clark oxygen electrodes, electrically insulated from the bathing solutions by Teflon or polyethylene membranes. Other details of methods are given elsewhere (Vieira et al., 1972*a*).

RESULTS AND DISCUSSION

Oxygen Uptake from the Outer and Inner Solutions

The skins of *Rana pipiens* and of *Bufo marinus ictericus* always took up less oxygen from the inner solution than from the outer solution. For short-circuited skins the uptake ratios had a mean value of 0.57 ± 0.09 (SE) for *Rana pipiens* (Vieira et al., 1972*a*) and 0.46 ± 0.11 (SE) for *Bufo marinus ictericus* (Danisi and Lacaz Vieira, *unpublished observations*). Reasons for these different rates of uptake may vary. However, it seems probable that the mitochondria-rich epithelial cells will consume oxygen more rapidly than the connective tissue, which will also function *in vitro* as a diffusion barrier for oxygen from the inner solution. These results are in agreement with those of Martin and Diamond (1966) in the rabbit gallbladder and of Nellans and Finn (1970) in the toad bladder.

Coupling Between Active Sodium Transport and Rate of Oxygen Consumption

Stoichiometric Ratio

When the rates of sodium transport (evaluated by short-circuit current) and of oxygen consumption are measured simultaneously with NaCl-Ringer's solution on both sides of skin, the spontaneous decline of sodium transport and of the associated rate of oxygen consumption enables the calculation of the stoichiometric ratio ($\Delta J^a_{Na}/\Delta J^*_{ro}$) (ions Na/O_2 molecule). Vieira et al. (1972*a*) and Danisi and Lacaz Vieira (1974) have shown for *Rana pipiens* and *Bufo marinus ictericus*, respectively, that in the whole range of sodium transport studied (from zero to approximately 150 μA cm^{-2}) the stoichiometric ratio is constant for a given skin. This permits the evaluation, by extrapolation, of the basal rate of oxygen consumption, which is the rate of oxygen consumption not directly associated with the trans-epithelial sodium transport. This quantity is about one half the total rate of oxygen consumption in a representative freshly mounted skin of *Rana pipiens* (Vieira et al., 1972*a*) or *Bufo marinus ictericus* (Danisi and Lacaz Vieira, *unpublished observations*). On the other hand, the various skins show significantly different values. Although the reasons for these variations are unknown, certain possibilities come to mind. One is recirculation of actively transported sodium (Ussing, 1966). If the magnitude of such recirculation were to differ from one skin to the other (possibly due, among other reasons, to different conditions of the skin in the molting cycle), the apparent stoichiometric ratio would vary. In this case a different stoichiometric ratio could possibly be a consequence of different degrees of sodium leakage from the interspace to the outer solution. Skin thickness and metabolic differences could also play an important role in the value of the stoichiometric ratio.

Action of Drugs on the Stoichiometric Ratio

Pitressin

This hormone increases the rate of sodium transport and the associated rate of oxygen consumption in *Rana pipiens* (NaCl-Ringer's solution on both sides of skin, and $\Delta\psi = 0$) (Vieira et al., 1972*a*). The effect has also been reported elsewhere in the literature (Ussing and Zerahn, 1951; Zerahn, 1956; Leaf and Renshaw, 1957). The stoichiometric ratio under the effect of pitressin is not significantly different from that observed during spontaneous decline of sodium transport (Vieira et al., 1972*a*).

Ouabain

Ouabain inhibits the rate of sodium transport to progressively lower values (Koefoed-Johnsen, 1957; Levy and Richards, 1965). The stoichiometric ratio during the spontaneous decline of sodium transport (15.0 ± 2.0 (SE) Na ions/O_2 molecule) was not statistically different from the value after addition of ouabain (13.8 ± 1.5 (SE) Na ions/O_2 molecule) ($0.60 < p < 0.70$). Similarly, the basal rate of oxygen consumption calculated by extrapolation $(J_{ro})_{I_o=0}$ under these two conditions (25.3 ± 8.8 (SE) pmole $cm^{-2}sec^{-1}$ and 28.4 ± 4.4 (SE) pmole $cm^{-2}sec^{-1}$) was not statistically different ($0.50 < p < 0.60$) for *Bufo marinus ictericus* (Danisi and Lacaz Vieira, 1974). The finding of a constant stoichiometric ratio is compatible with ouabain blocking a progressively greater number of apparent transport units with time, without altering their intrinsic stoichiometric ratio. The equality between $(J_{ro})_{I_o=0}$ and the rate of oxygen consumption after complete inhibition of J_{Na}^a by ouabain supports the idea that the basal rate of oxygen consumption is a parameter not directly related to sodium transport.

2,4-Dinitrophenol

The inhibitor 2,4-dinitrophenol (DNP) uncouples sodium transport and oxygen consumption (Fuhrman, 1952; Huf, Doss and Wills, 1957). It progressively reduces the transport to very small values. Simultaneously the rate of oxygen consumption increases up to four or five times the initial rate, being followed by a rapid subsequent decline (Vieira et al., 1972a). A similar transitory behavior is observed in the basal rate of oxygen consumption when the suprabasal rate of oxygen consumption is previously abolished by ouabain. The effect presumably reflects DNP's action on mitochondrial oxidative phosphorylation.

Cyanide

KCN not only inhibits active sodium transport but also abolishes total rate of oxygen consumption. In toad skin (*Bufo marinus ictericus*), KCN blocks completely the total rate of oxygen consumption in less than 2 min, and the short-circuit current obtained after that time declines exponentially with time at least for 60 min (Danisi and Lacaz Vieira, 1974). Figure 1 shows two representative experiments. If it is assumed that after 2 min there is a complete block in aerobic ATP production and also that this is the main ATP source, then we could write:

$$\frac{d(\text{ATP})}{dt} = -k\,(\text{ATP}) \tag{8}$$

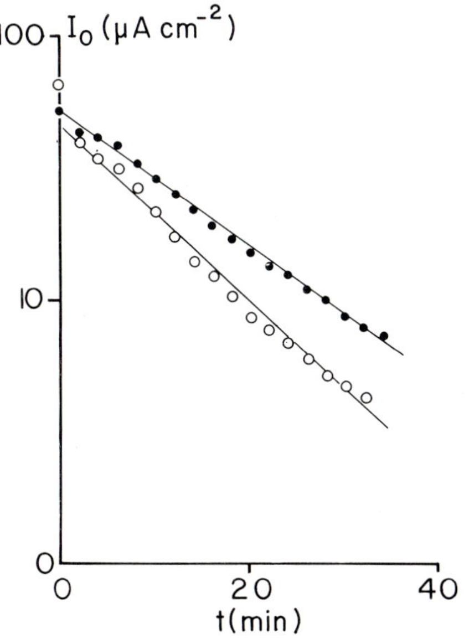

FIG. 1. Exponential decline of short-circuit current (I_o) with time. The values are those obtained 2 min after addition of KCN (1 mM) to both internal and external solutions. Two representative experiments with the skin of *Bufo marinus ictericus*.

$$I_o = -k_1 \frac{d(\text{ATP})}{dt} = k\, k_1\, (\text{ATP}) \tag{9}$$

Integrating Eq. 8,

$$(\text{ATP}) = (\text{ATP})_{t=0} e^{-kt} \tag{10}$$

Therefore,

$$I_o = k\, k_1\, (\text{ATP})_{t=0} e^{-kt} \tag{11}$$

The logarithmic dependence of I_o on time ($\ln I_o = a + bt$) strongly suggests the possibility that the transport system works at the expense of an ATP pool that is progressively depleted with time, and that the pool is apparently not compartmentalized within the cells, since its decay follows a one-compartment kinetics. These results are consistent with a minor role of anaerobic metabolism in sustaining sodium transport in toad skin.

Influence of Na Electrochemical Potential Difference on the Rates of Active Na Transport and Oxygen Consumption

According to Essig and Caplan (1968), the rate of active sodium transport and its associated rate of oxygen consumption could be described by phenomenological equations of nonequilibrium thermodynamics (Eqs. 1 and 2).

In freshly mounted skins of *Rana pipiens,* changes in $\Delta\psi$ resulted promptly in changes in the rate of oxygen consumption, normally within less than 30 sec, and usually a new steady-state value was reached within less than 2 min (Vieira, Caplan, and Essig, 1972*b*). Figure 2 shows a representative experiment of voltage clamping. Positive perturbations of $\Delta\psi$, which decreased the rate of sodium transport, reduced the rate of oxygen consumption and vice versa. When similar experiments are performed in the presence

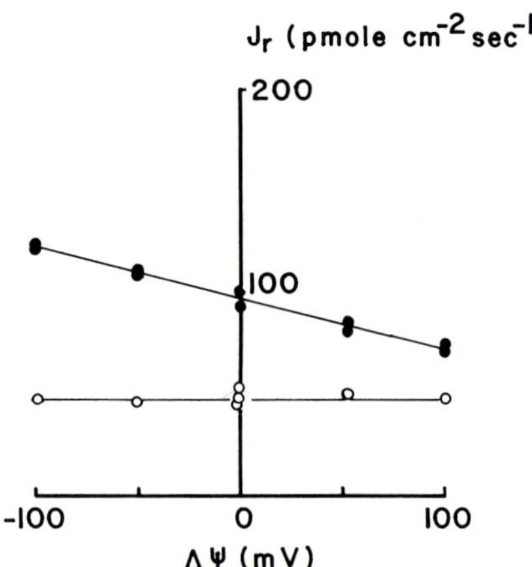

FIG. 2. Dependence of the rate of oxygen consumption on the electrical potential difference across the skin. Freshly mounted skin of *Rana pipiens. Filled circles:* control condition; *open circles:* after complete block of Na transport by ouabain (1 mM).

of ouabain (after a complete block of sodium transport) the remaining basal rate of oxygen consumption is no longer a function of $\Delta\psi$, as shown in Fig. 2. This fact demonstrates that the influence of $\Delta\psi$ on the rate of oxygen consumption is mediated through an effect on the active sodium transport.

In aldosterone-treated skins (10^{-6} M of d-aldosterone-21-acetate), which presented highly stable values of short-circuit current after 16 hr of incubation, a linearity was demonstrated between the rate of oxygen consumption and $\Delta\psi$ in the range of ±160 mV. Outside this voltage range, saturation of J_r as a function of $\Delta\psi$ was noticed, and J_r values obtained in the saturation range by positive clamping (internal solution positive) were comparable to the rates of basal oxygen consumption.

The dependence of J_r on $\Delta\psi$ permits the evaluation of the affinity of the metabolic driving reaction (A) (Essig and Caplan, 1968; Vieira et al., 1972b; Saito et al., 1973) according to the relation:

$$A = \frac{I_o}{\Delta J_r / \Delta(\Delta\psi)_A}. \tag{12}$$

For *Rana pipiens,* mean affinity values of 44.2 ± 13.2 (SE) kcal/mole of oxygen for freshly mounted skins and 107.1 ± 51.5 (SE) kcal/mole of oxygen for aldosterone-treated skins were obtained and could be compared with the value of some 116 kcal/mole of oxygen cited for glucose oxidation under physiological conditions (Davies and Ogston, 1950). Saito et al. (1973), studying the effect of aldosterone on the energetics of sodium transport in frog skin, were able to show that the hormone increases the affinity, as calculated by Eq. 12 under the condition of linearity.

Danisi and Lacaz Vieira (1974), applying the formalism of nonequilibrium thermodynamics (Essig and Caplan, 1968), were able to show for *Bufo marinus ictericus* that the rate of active sodium transport and the suprabasal rate of oxygen consumption are linear functions of the Na chemical potential difference across the short-circuited skin, in the range of 5 to 115 mEq/liter of external sodium, and that a stoichiometric ratio (L_{Na}/L_{rNa}) of 16.9 ± 0.8 (SE) Na ions/O_2 molecule was obtained in these experiments. This value is not statistically different from the stoichiometric ratio (L_{Nar}/L_r) calculated with equal sodium concentrations on both sides during spontaneous decline ($0.70 < p < 0.80$). The coefficient L_{Na} was estimated in long-term and short-term experiments with mean values of 0.28 ± 0.03 (SE) peq^{2}cal^{-1}cm^{-2}sec^{-1} and 0.45 ± 0.06 (SE) peq^{2}cal^{-1}cm^{-2}sec^{-1}, respectively, according to Eq. 4. This difference could be an indication of a decline of the phenomenological coefficients with time. The other coefficients were also estimated: $L_{Nar} = L_{rNa}$ (by Onsager's reciprocal relationship) equal to $1.7 \times 10^{-2} \pm 0.2 \times 10^{-2}$ (SE) peq mole cal^{-1} cm^{-2} sec^{-1} and $L_r = 1.1 \times 10^{-3}$ pmole^2 cal^{-1} cm^{-2} sec^{-1}. Própio and Lacaz Vieira (1974) have shown for *Bufo marinus ictericus* that in the external Na concentration range of 0.2 to 5.0 mEq/liter, the net Na flux (measured by chemical determination of this ion in the external solution), in non-voltage clamped skins, is a linear function of the electro-chemical potential difference across the skin. In their experiments, the electrical potential difference across the skin varied and was a logarithmic function of the external sodium concentration. A phenomenological coefficient, L_{Na}, calculated with their results, had a value of 0.38 pEq^2 cal^{-1} cm^{-2} sec^{-1}. That is comparable to the results of Danisi and Lacaz Vieira (1974) for the same parameter obtained under different experimental conditions in the same species.

Mandel and Curran (1973) have shown in *Rana pipiens* that active transport displays saturation as a function of applied potential and that both the level of saturation and the potential at which it is achieved are functions of the sodium concentration in the external solution. Under the assumption of a highly coupled system, the results of Mandel and Curran (1973) and of Danisi and Lacaz Vieira (1974) are in an apparent and indirect disagreement, considering the linearity of J_r^* with $\Delta\psi$ observed by the first authors and the saturation of the active transport with $\Delta\psi$ by the latter. Reasons for this discrepancy are still unknown.

The "Memory" Effect

Vieira et al. (1972*b*) described in *Rana pipiens* a phenomenon characterized by the fact that brief perturbations of $\Delta\psi$ ($\pm$ 30 to 200 mV for a few seconds) did not alter the subsequent values of the short-circuit current (I_o) and the rate of oxygen consumption (J_{ro}). Longer perturbations produced a memory effect: positive clamping, which slowed active sodium transport, transiently increased the subsequent values of I_o and of J_{ro}; negative clamping induced the opposite effect. Danisi and Lacaz Vieira (1974) extended the observations on the memory effect for *Bufo marinus ictericus,* showing that a block of sodium transport (by removal of sodium from the external solution) is followed by a subsequent increase in J_{Na}^a and J_{ro} (when sodium is returned to the external solution) to values higher than the ones observed before block. The ratio L_{Nar}/L_r is maintained after blocking sodium transport. Their results also show that the effect in sodium transport is due to an increase in the product $L_{Nar}A,$ without a significant alteration of L_{Na}. It would be an important point to know whether this increase is due to changes in L_{Nar} or in A or in both. On a speculative basis, it is most probable that the increase is due to an increase in A, which could result from an accumulation of metabolic intermediates—probably not ATP, since in terms of the oxidative phosphorylation regulatory mechanism, an increase in an ATP pool related to transport (after sodium-transport block) would result in an increase in the rate of sodium transport, although accompanied by a reduction in the rate of oxygen consumption. One possible explanation might be, for example, that the accumulation of ATP causes an increase in the NADH/NAD ratio and with the rapid lowering of ATP concentration following the release of block, the high NADH/NAD ratio then causes rapid oxygen consumption.

ACKNOWLEDGMENTS

This work was supported by a grant from the São Paulo State Research Foundation (FAPESP).

REFERENCES

Conway, E. J. (1951): The biological performance of osmotic work. A redox pump. *Science* 113:270–273.

Conway, E. J. (1953): A redox pump for the biological performance of osmotic work, and its relation to the kinetics of free ion diffusion across membranes. *Int. Rev. Cytol.* 2:419–445.

Conway, E. J. (1955): Evidence for a redox pump in active transport of cations. *Int. Rev. Cytol.* 4:377–396.

Danisi, G., and Lacaz Vieira, F. (1974): Nonequilibrium thermodynamic analysis of the coupling between active sodium transport and oxygen consumption. *J. Gen. Physiol.* 64:372–391.

Davies, R. E., and Ogston, A. G. (1950): On the mechanism of secretion of ions by gastric mucosa and by other tissues. *Biochem. J.* 46:324–333.

Dean, R. B., and Gatty, O. (1937): The bioelectric properties of frog skin. *Trans. Faraday Soc.* 33:1040–1046.

Essig, A., and Caplan, S. R. (1968): Energetics of active transport processes. *Biophys. J.* 8:1434–1457.

Fuhrman, F. A. (1952): Inhibition of active sodium transport in the isolated frog skin. *Am. J. Physiol.* 171:266–278.

Huf, E. G. (1955): Ion transport and ion exchange in frog skin. In: *Electrolytes in Biological Systems,* edited by A. M. Shanes, pp. 205–238. American Physiological Society, Washington, D.C.

Huf, E. G., Doss, N. S., and Wills, J. P. (1957): Effects of metabolic inhibitors and drugs on ion transport and oxygen consumption in isolated frog skin. *J. Gen. Physiol.* 41:397–417.

Koefoed-Johnsen, V. (1957): The effect of strophanthin (ouabain) on the active transport of sodium through the isolated frog skin. *Acta Physiol. Scand.* 42:87, Suppl. 145.

Leaf, A., and Renshaw, A. (1957): Ion transport and respiration of isolated frog skin. *Biochem. J.* 65:82–90.

Levy, J. V., and Richards, V. (1965): Action of aldosterone and ouabain on oxygen consumption and short-circuit current of toad bladder. *Proc. Soc. Exp. Biol. Med.* 118:501–507.

Linderholm, H. (1952): Active transport of ions through frog skin with special reference to the action of certain diuretics. *Acta Physiol. Scand.* 27:1, Suppl. 97.

Mandel, L. J., and Curran, P. F. (1973): Response of the frog skin to steady-state voltage clamping. II. The active pathway. *J. Gen. Physiol.* 62:1–24.

Martin, D. W., and Diamond, J. M. (1966): Energetics of coupled active transport of sodium and chloride. *J. Gen. Physiol.* 50:295–315.

Nellans, H., and Finn, A. (1970): Energetics of sodium transport in the toad bladder. *Biophys. Soc. Ann. Meet. Abstr.* 10:196a.

Prócopio, J., and Lacaz Vieira, F. (1974): Net transport of ions in the isolated toad skin under low ionic concentration in the external solution. (*ms. in preparation*).

Saito, T., Essig, A., and Caplan, S. R. (1973): The effect of aldosterone on the energetics of sodium transport in the frog skin. *Biochim. Biophys. Acta* 318:371–382.

Ussing, H. H. (1953): Transport through biological membranes. *Annu. Rev. Physiol.* 15:1–20.

Ussing, H. H. (1954): Ion transport across biological membranes. In: *Ion Transport across Membranes,* edited by H. T. Clarke, pp. 3–22. Academic Press, New York.

Ussing, H. H. (1960): *The alkali metal ions in biology. Handb. Exptl. Pharmakol.* Springer-Verlag, Berlin, vol. XIII, 1.

Ussing, H. H. (1966): Anomalous transport of electrolytes and sucrose through the isolated frog skin induced by hypertonicity on the outside bathing solution. *Ann. N.Y. Acad. Sci.* 137:543–555.

Ussing, H. H., Erlij, D., and Lassen, U. (1974): Transport pathways in biological membranes. *Annu. Rev. Physiol.* 36:17–49.

Ussing, H. H., and Zerahn, K. (1951): Active transport of sodium as the source of electric current in the short-circuited isolated frog skin. *Acta Physiol. Scand.* 23:110–127.

Vieira, F. L., Caplan, S. R., and Essig, A. (1972a): Energetics of sodium transport in frog skin. I. Oxygen consumption in the short-circuited state. *J. Gen. Physiol.* 59:60–76.

Vieira, F. L., Caplan, S. R., and Essig, A. (1972b): Energetics of sodium transport in frog skin. II. The effects of electrical potential on oxygen consumption. *J. Gen. Physiol.* 59:77–91.

Zerahn, K. (1956): Oxygen consumption and active sodium transport in the isolated short-circuited frog skin. *Acta Physiol. Scand.* 36:300–318.

Zerahn, K. (1958): Oxygen consumption and active sodium transport in isolated amphibian skin under varying experimental conditions. *Thesis.* Universitetsforlaget I Aarhus, Aarhus, Denmark.

Hydrogen Ion Transport in Renal Tubules: Effect of Acetazolamide

G. Malnic, Margarida de Mello Aires, and A. C. Cassola

*Department of Physiology, Institute of Biomedical Sciences, University of São Paulo,
01000 São Paulo, S. P., Brazil*

INTRODUCTION

The relation between the enzyme carbonic anhydrase and acid secretion in several epithelia has long been recognized. The development of specific inhibitors of this enzyme, such as acetazolamide and its analogues, was consequently of great value for investigation of the mechanisms of acid formation and secretion in such structures as gastric mucosa, turtle bladder, and kidney. Early studies in kidney showed that acetazolamide administration leads to alkalinization of the urine because of increased bicarbonate excretion (Berliner, Kennedy, and Orloff, 1951; Maren, Wadsworth, Yale, and Alonso, 1954). Furthermore, Maren (1967) showed that at a dosage of 5 to 20 mg/kg body weight, acetazolamide leads to effectively complete inhibition of carbonic anhydrase, greatly enhancing acetazolamide's usefulness for studies on the mechanism of action of carbonic anhydrase and of H ion secretion.

Biochemical studies have shown that carbonic anhydrase occurs predominantly in the supernatant phase of kidney cell homogenates, that is, in cytoplasm. A sizable fraction, however, is bound to microsomes (Maren and Ellison, 1967). The fractions present some differences in kinetic behavior, but both are reported to be completely inhibited by dosages attainable *in vivo*. However, the particulate fraction needs significantly higher drug levels to attain total inhibition. These findings are compatible with physiological and histochemical evidence indicating that carbonic anhydrase, besides occurring in considerable concentrations in tubular cells, attains significant levels on the brush border of the proximal tubule. This was shown histochemically by Hansson (1968) and by Loennerholm (1973). In the distal tubule, on the other hand, the enzyme is found in the cytoplasm but not on the luminal border.

DISEQUILIBRIUM pH IN RENAL TUBULES

Histochemical and biochemical data on the renal distribution of carbonic anhydrase agree with physiological findings that a luminal disequilibrium pH is found in the distal but not in the proximal tubule of normal rats. Upon inhibition of carbonic anhydrase, a considerable disequilibrium pH is found in both segments (Rector, Carter, and Seldin, 1965; Vieira and Malnic, 1968). The origin of a disequilibrium pH can be visualized with the help of the following relations:

$$CO_2 + H_2O \underset{}{\overset{c.a.}{\rightleftharpoons}} H_2CO_3 \rightleftharpoons H^+ + HCO_3^- \tag{1}$$

where the first step, the hydration of CO_2 or dehydration of H_2CO_3, is catalyzed by carbonic anhydrase (c.a.);

$$pH = pK + \log \frac{HCO_3^-}{CO_2 + H_2CO_3} \tag{2}$$

the Henderson-Hasselbalch equation, where, in the case of the bicarbonate buffer system, the acid term includes conventionally both CO_2 and H_2CO_3, and where $pK = 6.1$;

$$\dot{V}_{CO_2} = k_{CO_2} \cdot (CO_2) \tag{3}$$

where $\dot{V}_{CO_2}$ is the rate of spontaneous hydration of CO_2, without enzyme catalysis, k_{CO_2} is the hydration rate coefficient, with a value of $0.15 \ \text{sec}^{-1}$ at 38°C (according to Magid and Turbeck, 1968; and Garg and Maren, 1972), and (CO_2) is the concentration of carbon dioxide. The reverse reaction, carbonic acid dehydration, has a similar form but includes the dehydration rate coefficient $k_{H_2CO_3}$ (of the order of 50 sec^{-1}) and H_2CO_3 concentration:

$$\dot{V}_{H2CO3} = k_{H2CO3} \cdot (H_2CO_3) \tag{4}$$

When acidification of tubular fluid proceeds via H ion secretion, H_2CO_3 is formed in the lumen and further dehydrated there to CO_2 and water. The latter step may be rate limiting in the absence of carbonic anhydrase or after its inhibition. In this case, in order for the dehydration reaction to proceed at the same rate as H ion secretion, luminal carbonic acid concentration will be increased. Enzyme presence increases the rate coefficients by approximately 10,000; in the enzyme's absence, the rate of reactions 3 and 4 can be increased only by increasing substrate concentration.

Equation 2, on the other hand, describes the pH of the buffer system under equilibrium conditions, where the proportion of CO_2 to H_2CO_3, given by the relation k_{H2CO3}/k_{CO2}, is of the order of 330. When luminal H_2CO_3 concentration increases because of absence or inhibition of carbonic anhydrase, the luminal pH will be more acid than expected from the ambient pCO_2 and bicarbonate concentration. The difference between this pH and that measured in a droplet of collected tubular fluid equilibrated with the animal's pCO_2 is the disequilibrium pH. Experimentally, a disequilibrium pH of 0.88 units was found by Rector et al. (1965) in the distal tubule of rats with moderate bicarbonate infusion, whereas Vieira and Malnic (1968) found a disequilibrium pH of 0.34 in control rats. In the proximal tubule, a disequilibrium pH was found only after carbonic anhydrase inhibition. A value of 0.85 pH units was found by Rector et al. (1965) and a value of 0.41 was found by Vieira and Malnic (1968). These disequilibrium pH values indicate that bicarbonate reabsorption proceeds, at least in part, by H ion secretion and not by bicarbonate reabsorption as such.

On the other hand, based on acetazolamide's degree of inhibition of bicarbonate reabsorption in dogs, Maren (1967) proposed that approximately 60% of bicarbonate reabsorption should be the result of ionic bicarbonate transfer, because that proportion is still reabsorbed after enzyme inhibition. Maren maintains that acetazolamide and similar drugs should inhibit most H ion secretion based on CO_2 hydration; the remaining fraction would be the result in part of noncatalyzed CO_2 hydration—corresponding to 10 to 15% of the total acidification rate—and to ionic bicarbonate reabsorption. However, the occurrence of a sizable disequilibrium pH after acetazolamide administration should favor the existence of continued H ion secretion even under these circumstances. Another experimental finding that supports this view is the stimulation of bicarbonate reabsorption by increased blood pCO_2 after carbonic anhydrase inhibition, indicating that transport depends on the cellular CO_2 level—that is, on CO_2 hydration (Rector, Seldin, Roberts, and Smith, 1960).

SEGMENTAL RATES OF BICARBONATE REABSORPTION

Bicarbonate reabsorption along individual nephron segments can be calculated from bicarbonate concentrations and changes in inulin concentration ratios along these segments. In Fig. 1, the relation between filtered load and bicarbonate reabsorption in the proximal tubule is plotted for a series of different acid-base situations. In control rats and in metabolic acidosis, almost all filtered bicarbonate is reabsorbed in this segment. The reabsorption is markedly decreased in various alkalotic conditions, but the greatest reduction in fractional reabsorption is found after acetazolamide treatment. Treatment by acetazolamide is one of the few conditions, besides osmotic diuresis, in which a significant reduction in fractional volume reabsorption is found in the proximal tubule. In acetazolamide-treated rats, a significant fall in inulin TF/P from a control value of 2.41 to 1.69 for locations between 40 and 65% of tubular length was found (Malnic, Mello Aires, and Vieira, 1970). A depression of proximal bicarbonate and fluid reabsorption was also observed by Clapp, Watson, and Berliner (1963) and by Beck and Goldberg (1973) in recollection micropuncture experiments.

It is of interest that not only bicarbonate but also chloride reabsorption is reduced, as can be observed in both free-flow and stopped-flow microperfusion studies (Malnic and Mello Aires, 1970; Kunau, 1972). Actually, an effect of acetazolamide on chloride transport in different biological mem-

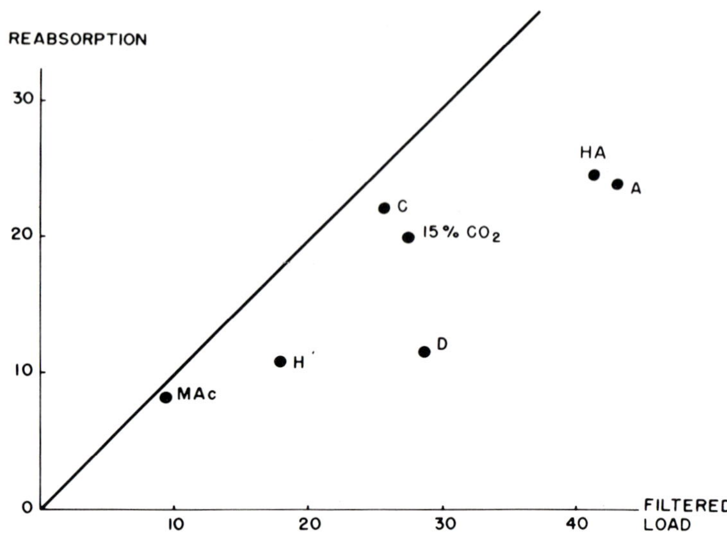

FIG. 1. Relation between filtered load of bicarbonate and its reabsorption along the proximal tubule (μEq/min · kg/ml GFR). The line represents complete reabsorption of the load. C, control; A, alkalosis; HA, hypochloremic alkalosis; D, diamox; H, hyperventilation; MAC, metabolic acidosis; 15% CO_2, respiratory acidosis. Data from Malnic, Mello Aires, and Giebisch (1972), and Mello Aires and Malnic (1972).

brane systems has been observed. Thus, for instance, Durbin and Heinz (1958) observed inhibition of chloride transport not coupled to H ion transport in frog gastric mucosa; and Maren and Broder (1970) observed a similar inhibition of chloride transfer into cerebrospinal fluid. The relation between carbonic anhydrase and chloride transport, however, is not clear. An indirect action via osmotic diuresis or interaction with changes in volume flow does not appear to explain all experimental observations. It is possible that acetazolamide might act on a specific chloride-transfer system independent of carbonic anhydrase. But according to NMR studies by Ward (1969), there appears to be an interaction between chloride and carbonic anhydrase, suggesting that the enzyme might be, or participate in, a transport system for the ion.

An analysis of bicarbonate reabsorption in the distal tubule, as summarized in Table 1, shows that the highest fractional reabsorption is found in the control group. In absolute terms, however, bicarbonate reabsorption is markedly increased in both acute and chronic alkalosis and after acetazolamide infusion, probably because of the considerable increase in the segmental bicarbonate load. On the other hand, in acidotic conditions, net distal bicarbonate reabsorption is both fractionally and absolutely reduced, perhaps in part because of the decreased distal load of this ion. These data show that it is difficult to evaluate the actual acidifying capacity of the tubular epithelium by free-flow methods—especially in the case of the distal tubule—because net reabsorption rates depend to a considerable extent on the bicarbonate load of the particular tubular segment.

TABLE 1. *Bicarbonate load and reabsorption in distal tubule under different experimental conditions ($\mu Eq/min \cdot kg/ml$ GFR)*

Condition	Load	Reabsorption	R/L
Control	3.1	2.28	0.74
Acetazolamide	16.7	9.30	0.56
Acute alkalosis	16.0	7.02	0.44
Hypochloremic alkalosis	8.3	4.37	0.53
Hyperventilation	3.9	2.05	0.53
15% CO_2	1.7	1.11	0.65
Metabolic acidosis	0.88	0.40	0.45

R/L: reabsorption/load. Data from Malnic, Mello Aires, and Giebisch (1972) and Mello Aires and Malnic (1972).

STOPPED-FLOW MICROPERFUSION STUDIES

We have recently developed a method that permits the evaluation of tubular acidification in cortical nephron segments independent of the filtered load of buffer (Malnic and Mello Aires, 1971). This method is based on the intraluminal injection of alkaline or acid buffer systems, isolating the perfusion fluid between castor oil columns, and recording subsequent pH changes by means of an antimony microelectrode impaled in the perfusate column. In general, 100 mM bicarbonate or alkaline phosphate, brought to isotonicity with raffinose, is used. The antimony microelectrode—despite its sensitivity to interfering substances like proteins and oxidizing and reducing compounds, and occasional discrepant pH readings when buffers of different anionic content are used (Karlmark and Sohtell, 1973)—is well suited to continuous tubular pH recordings because of its rapid response to pH changes (Malnic and Vieira, 1972).

A schematic drawing of pH changes during tubular perfusion with an alkaline phosphate solution is shown in Fig. 2. Perfusion is performed by means of a double-barreled micropipette for injection of perfusion fluid and colored castor oil. On starting perfusion, the recorded pH rises rapidly to the alkaline pH of the buffer solution. Immediately after the tubule becomes blocked with oil, the pH starts to fall, tending exponentially to its equilibrium level, reached after some 30 to 60 sec under normal conditions. Figure 2 also shows an acidification curve found after carbonic anhydrase inhibition; it is characterized by a more alkaline steady-state level and a slower approach to this level.

Similar curves can be obtained during perfusion with bicarbonate solu-

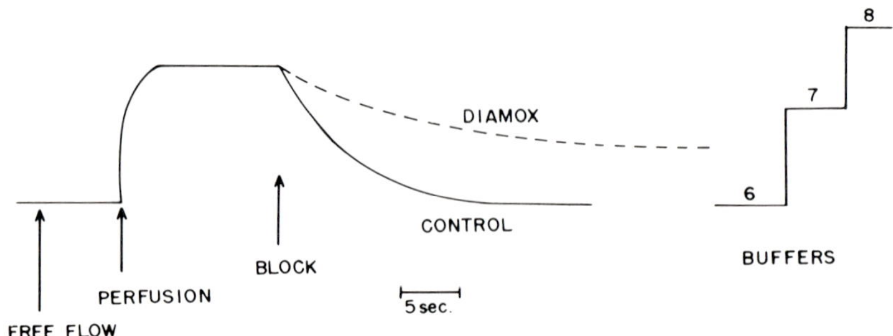

FIG. 2. Proximal tubular acidification curves during perfusion with alkaline phosphate in control and diamox-infused rats, obtained with antimony microelectrodes. *Right side,* standardization of electrodes with buffer solutions.

tions. When bicarbonate solutions are preequilibrated with a pCO$_2$ near that of the experimental animal, bicarbonate concentrations can be calculated from the instantaneous pH and the animal's pCO$_2$ by the Henderson-Hasselbalch equation. These concentrations approach their steady-state levels exponentially; thus a plot of the logarithm of the difference between bicarbonate concentrations at time t and at steady state against time in seconds yields a straight line, as shown in Fig. 3. In acetazolamide-treated animals, a similar line is obtained, but its slope is decreased. From these graphs, half-times of the approach of bicarbonate concentrations to their steady-state levels can be obtained. Note that after drug treatment half-time is markedly increased.

When tubules are perfused with air-preequilibrated solutions, however, the first period of perfusion yields higher bicarbonate concentrations than the values given in Fig. 3, because at that time tubular pCO$_2$ is still lower than the plasma level. Only after a period of some 5 sec is CO$_2$ equilibrium between perfusion fluid and blood reached, after which the regular exponential concentration decay is found. This line can be extrapolated back to zero in order to obtain the actual bicarbonate concentrations for the initial perfusion period; from these values and the instantaneous pH, carbonic acid concentrations can be calculated using the Henderson-Hasselbalch equation but introducing the pK corresponding to the dissociation of this acid; that is, 3.57, according to Gibbons and Edsall (1963). When the carbonic acid concentrations are plotted against time, as shown in Fig. 4, the rate of approach of these levels to plasma carbonic acid concentrations can be observed. In control rats, this approach has a half-time of 0.5 to 2 sec and the appearance of a simple exponential. After carbonic anhydrase inhibition, however, approach to plasma levels is considerably retarded, corresponding to a mean half-time of 4.5 sec. Since reaction kinetics show that this retarda-

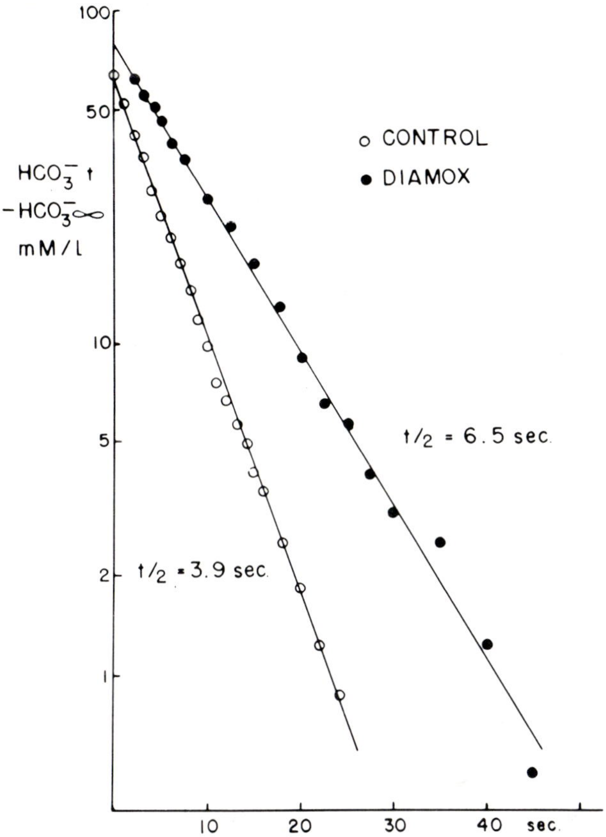

FIG. 3. Fall of bicarbonate concentrations in the proximal tubule of control and diamox-infused rats. Concentrations were calculated from acidification curves obtained during perfusion with 100 mM NaHCO$_3$. From Malnic and Giebisch (1972a).

tion cannot be the result of delayed hydration of CO_2 in the tubular lumen, it could be caused by delayed CO_2 inflow from blood to lumen (Malnic and Mello Aires, 1971).

The absolute value of the CO_2 equilibration half-times may be affected by the intrinsic response time of the electrode—which has a half-time of a few tenths of a second—or by the diffusion into the lumen in that brief period of other, non-bicarbonate buffers, which would be titrated by the carbonic acid formed. In any case, the observed values represent a lower limit for CO_2 permeation through the tubular epithelium and have been used to evaluate luminal pCO_2 levels in control and acetazolamide-infused rats (Malnic, Mello Aires, and Cassola, 1974a). Such calculations lead to the possibility that part of the measured disequilibrium pH, especially after carbonic anhydrase inhibition, might be the result of increased luminal CO_2

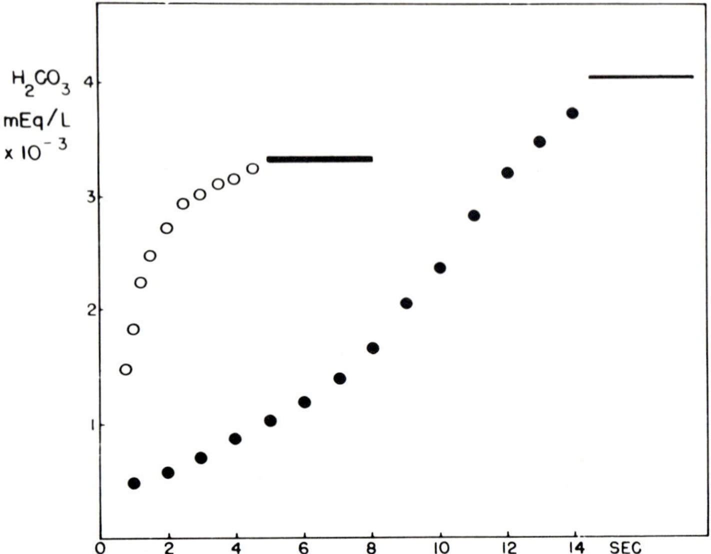

FIG. 4. Approach of carbonic acid concentrations in proximal tubule to plasma levels during perfusion with bicarbonate solution in control (○) and acetazolamide-infused (●) rats. From Malnic and Mello Aires (1971).

levels (see Eq. 2), as previously suggested by Karlmark (1973). However, more direct measurements of luminal pCO$_2$ *in vivo* will be necessary to actually demonstrate the occurrence of a disequilibrium pCO$_2$ at this site.

INTERPRETATION OF ACIDIFICATION KINETICS

The data obtained in stopped-flow microperfusion experiments of the type described above can be used to analyze acetazolamide's mode of action on the acidification process in renal tubules. For this purpose, the measured changes in pH have to be interpreted in terms of a kinetic model. The simplest would be a two-compartment system: compartment 1 being the tubular lumen, where pH is measured; and compartment 2 the tubular cell, or extraluminal space in general—the cell and peritubular space as well as capillaries being considered in a steady state not affected by the experimental procedure itself, that is, by tubular microperfusion.

Thus the changes in luminal acid content, S_1, would be given by:

$$\frac{dS_1}{dt} = -k_{12}S_1 + k_{21}S_2, \tag{5}$$

where S_1 and S_2 are acid concentrations for a fixed segment of tubular lumen (S_1) or cell (S_2), and the k's are rate coefficients corresponding to H ion flow from compartment 1 to compartment 2 or vice versa. Acid content is the level of acid phosphate during perfusion with the buffer, i.e., the total phosphate minus alkaline phosphate. By analogy, if the mechanism of bicarbonate reabsorption is accepted, then during bicarbonate perfusion, the acid content of tubular lumen is the initial bicarbonate concentration minus bicarbonate at time t, because the difference corresponds to the quantity of H ions secreted into the lumen.

According to this model, $k_{12}S_1$ is the flux of H ions from lumen to cell, and $k_{21}S_2$ the flux from cell to lumen. At steady state, the fluxes are equal and no change in S_1 is observed. In order to yield an exponential decay of the difference between the final acid concentration ($S_{1\infty}$) and the concentration at time t, a simplifying assumption has to be made for Eq. 5. Thus it could be assumed that the H ion secretory rate ($k_{21}S_2$) is constant, whereas the leakage of H ions from the lumen ($k_{12}S_1$) varies in magnitude according to the luminal acid content, S_1. In this case, solution of Eq. 5 yields one exponential, with a rate coefficient of k_{12} given by the slope of lines such as those shown in Fig. 3; its intercept at the ordinates is the steady-state acid content (or $HCO_3^-{}_0 - HCO_{3\infty}^-$). Based on this model, the rate coefficient obtained would be related to passive H ion permeability; and the unidirectional H ion flow from cell to lumen would be evaluated by $k_{12}S_{1\infty}$, the product of two experimentally obtained values (Malnic and Giebisch, 1972*a*, and Malnic, Mello Aires, de Mello, and Giebisch, 1975).

A summary of mean acidification parameters for proximal and distal tubule during phosphate and bicarbonate perfusions in control and aceta-

TABLE 2. *Acidification in renal tubule in control and acetazolamide-infused rats; comparison with Na reabsorption*

Group	Tubule	Perfusion	pH_∞	Half-time (sec)	ϕ_H mEq/liter·sec	ϕ_H nEq/cm²·sec	% Control
Control	Proximal	Ph	6.23	7.43	6.93	5.20	
		B	6.21	5.15	13.2	9.92	
	Distal	Ph	6.63	8.29	4.53	2.83	
		B	6.78	4.38	17.3	9.22	
Acetazolamide	Proximal	Ph	7.07	13.2	1.76	1.32	25%
		B	6.83	8.12	7.92	5.95	60%
	Distal	Ph	7.03	23.6	1.10	0.69	24%
		B	6.89	21.9	2.93	1.83	20%

Group	Tubule	Perfusion	ϕ_{Na}, nEq/cm²·sec	% Control
Control	Proximal	NaCl	9.17	
	Distal	NaCl	2.4[a]	
Acetazolamide	Proximal	NaCl	4.73	52%

Luminal perfusions: Ph: 100 mM Na_2HPO_4; B: 100 mM $NaHCO_3$; NaCl: 150 mM NaCl. Fluxes are given in terms of luminal concentration changes and per cm² tubular epithelium.

Data from Malnic et al. (1975) and Malnic and Mello Aires (1970).

[a] From Gertz (1963).

zolamide-infused rats is given in Table 2. Carbonic anhydrase inhibition was performed by giving a prime of 20 mg/kg of diamox, and thereupon starting an infusion of 20 mg/kg diamox/hr—a dose that, according to Maren (1969), leads to practically complete inhibition of carbonic anhydrase. Steady-state pH levels are markedly increased in acetazolamide-infused rats, showing that they have a decreased ability to maintain transtubular H ion gradients. Furthermore, acidification half-times are significantly increased in both proximal and distal tubule, but especially in the latter segment, indicating a reduced k_{12} (since $k = \ln 2/\ t/2$), that is, a reduced H ion permeability. From these values and the steady-state acid content calculated from the steady-state pH values, H ion fluxes from cell to lumen are calculated in terms of mEq/liter/sec and per cm^2 tubular wall.

NaCl TRANSPORT AND CARBONIC ANHYDRASE

In Table 2, NaCl transport rates are given in the units used to describe H ion secretion or bicarbonate reabsorption. Net NaCl fluxes were obtained by the Gertz (1963) split-droplet method. These fluxes correspond to about a third of the unidirectional fluxes and are generally thought to represent the active transfer component of the salt (Morel and Murayama, 1970). According to Froemter, Rumrich, and Ullrich (1973), however, only about a third of the net flux of Na is the result of active transport itself, the remainder being caused by passive movement such as diffusion and solvent drag. These net fluxes are of the order of the bicarbonate fluxes found in proximal tubule. It is of interest to note that both NaCl and bicarbonate reabsorption in proximal tubule are inhibited 40 to 50% by acetazolamide, indicating that NaCl transport is also somehow dependent on carbonic anhydrase. An alternative explanation would be a direct action of the drug on ion transport systems, independent of its action on carbonic anhydrase.

The work of Gonzalez (1969) on turtle bladder, showing independent action of acetazolamide on the short-circuit current related to chloride and bicarbonate transport as well as on unidirectional chloride fluxes, points toward drug action on a pump system or on a related driving reaction, common to both transport systems. Such action was also supported by Kitahara, Fox, and Hogben (1967), who found inhibition of chloride transport by acetazolamide in frog cornea, a structure supposedly devoid of carbonic anhydrase. Recently, however, carbonic anhydrase has also been demonstrated in that tissue (Loennerholm, 1974). At present, the relation between NaCl transport and carbonic anhydrase is not clear. The need of Na for H ion transport is apparently not absolute, since tubules perfused with choline bicarbonate acidify the urine to practically the same extent as controls (Malnic and Mello Aires, 1971). A possible participation of carbonic anhydrase in the NaCl-transport system cannot, however, be ruled out.

MECHANISM OF ACTION OF ACETAZOLAMIDE ON TUBULAR ACIDIFICATION

The extent to which acetazolamide inhibits acidification varies with the buffer solution used and the tubular segment studied. Inhibition with respect to controls is of the order of 75 to 80% in most situations, with the exception of bicarbonate perfusions in proximal tubule, where it is only 40%. There are several mechanisms by which this discrepancy might be explained. Because of the considerable passive permeability to ions of this segment— as evidenced by its low electrical resistance (Hegel, Froemter, and Wick, 1967)—chloride ions flow rapidly into the initially chloride-free fluid columns. This flow is accompanied by Na and water, leading to dilution of the luminal buffer (Malnic and Mello Aires, 1970) and participating in the reduction of bicarbonate concentration. Furthermore, participation of passive outflow of bicarbonate from the tubular lumen along a favorable concentration gradient has also to be considered. The latter fraction does not appear to be important, since proximal permeability to bicarbonate, despite being measurable, is considerably lower than that for chloride (Bank and Aynedjian, 1967; Froemter et al., 1973). Also, the acidification half-times obtained on perfusion with 100 and 25 mm bicarbonate are similar, showing that in the absence of a significant gradient for passive bicarbonate diffusion reabsorption of that anion proceeds with practically the same rate coefficient (Malnic and Mello Aires, 1971). According to Maren (1967), a sizable portion of bicarbonate reabsorption after carbonic anhydrase inhibition could be the result of bicarbonate transport as such, independent of H ion secretion, and proceeding by a mechanism supposedly insensitive to acetazolamide; the existence of an important fraction of bicarbonate transport by this mechanism will be discussed below.

According to the model of acidification adopted, acetazolamide appears to act on tubular permeability to H ions, inasmuch as k_{12} is markedly reduced in both proximal and distal tubule. Acetazolamide also reduces the secretory rate of H ions (ϕ_H) into the tubular lumen. A lower permeability at the same secretory rate would lead to increased transtubular H ion gradients, since the H ion leak would be decreased. Thus the decreased secretory flux leads to higher luminal steady-state pH despite the lower H ion permeability.

The proposed model for acidification of renal tubule is, however, not the only hypothetical system compatible with the experimental data. For turtle urinary bladder, Steinmetz and Lawson (1971) obtained data indicating that permeability to H ions in that structure was very low and not affected by acetazolamide, and that in the physiological range of acidification, H ion secretion varied inversely with mucosal H ion concentration. Thus acidification rate changed because of changes in pump activity and not because of variations in the back-leak of these ions. They found, further, that H ion permeability became significant below a mucosal pH of 4.4, where passive H ion flux from mucosa to serosa increased linearly with mucosal H ion concentration.

In order to test the pump-leak hypothesis more directly in our system, pH changes were measured after luminal perfusion with solutions containing 100 mM phosphate buffer at pH 5.5, that is, at a level more acid than steady-state pH. As during alkaline perfusions, luminal pH approached the steady-state level according to a single exponential. Alkalinization half-times obtained during such experiments in control and acetazolamide-infused rats are summarized in Table 3 and compared with control acidification half-time. In control proximal tubule, alkalinization and acidification half-times are of the same order of magnitude; that is compatible with the pump-leak hypothesis, according to which in both acid and alkaline per-fusions the rate coefficient measured experimentally is an evaluation of the passive permeability of the epithelium to H ions. After acetazolamide, alkalinization half-times increase markedly, in accordance with the findings during alkaline perfusions indicating that this drug effectively reduces H ion permeability in the proximal segment.

In the distal tubule, on the other hand, alkalinization half-times are significantly longer than acidification half-times in control rats, especially in early distal segments. In the latter, the rate coefficient obtained during acid perfusions is only about a third that obtained during alkaline perfusions. Thus the coefficient probably measuring passive permeability could be considered almost negligible compared with the total acidification coeffi-cient; such a situation could be compared to that in turtle bladder, charac-terized by low passive permeability and an H ion pump with activity in-versely proportional to the gradient against which it pumps.

These studies indicate, therefore, that proximal tubule and distal tubule behave in different ways with respect to H ion transport. That is com-patible with electrophysiological studies indicating that the proximal tubule

TABLE 3. *Mean acidification and alkalinization half-times in renal tubule of con-trol and acetazolamide-infused rats, perfused with alkaline or acid phosphate*

	Half-time (sec)		
Tubule	Acidification	Alkalinization	Alkalinization + Acetazolamide
Proximal	7.4	9.7[a]	23.3[b]
Distal	8.0	37.7[b]	31.8
Distal	8.6	16.9[b]	22.6[a]

[a] $(0.05 > p > 0.01)$.

[b] $(p < 0.01)$. Statistical comparisons of data in columns 2 and 1, and 3 and 2. Data from Cassola (1974).

is a rather leaky structure, with an electrical resistance of the order of 5 ohm · cm² (Hegel et al., 1967; Boulpaep and Seely, 1971) and considerable ionic permeation along intercellular spaces (Whittembury, Rawlins, and Boulpaep, 1973). The distal tubule, by contrast, has a considerably higher epithelial resistance — of the order of 300 ohm · cm² — and is therefore a much tighter structure, capable of maintaining considerable ionic and even osmotic gradients (Malnic and Giebisch, 1972*b*). Table 3 shows, furthermore, that in the distal tubule, acetazolamide does not appear to change epithelial permeability to H ions; despite a significant increase in half-times in late distal tubule, the observed alterations are minor compared to changes in the proximal segment. That would indicate that the decreased acidification in distal tubule after acetazolamide administration is caused by depression of the segment's capacity to secrete H ions without major alterations in H ion permeability.

SOURCE OF H IONS FOR TUBULAR SECRETION AFTER ACETAZOLAMIDE

The data obtained in the micropuncture experiments described permit a reassessment of the role of catalyzed and uncatalyzed CO_2 hydration and H_2CO_3 dehydration in acidification of renal tubule. Based on the kinetics of the uncatalyzed reaction, it is possible to calculate the expected luminal carbonic acid concentration after acetazolamide administration. According to Eq. 4, the rate of H_2CO_3 dehydration in tubular lumen can be obtained if one knows the dehydration rate coefficient and the luminal H_2CO_3 concentration. In a steady-state situation, the dehydration rate of H_2CO_3 should be equal to the rate of bicarbonate reabsorption, if all bicarbonate is reabsorbed via H ion secretion:

$$\dot{V}_{H_2CO_3} = \dot{V}_{HCO_3^-}$$

Therefore luminal H_2CO_3 concentration in steady state can be obtained by the relation:

$$(H_2CO_3) = \dot{V}_{HCO_3^-}/k_{H_2CO_3} \tag{6}$$

Using segmental reabsorption rates (Malnic, Mello Aires, and Giebisch, 1972) and the carbonic acid dehydration rate coefficient of 50 sec⁻¹, or 3,000 min⁻¹, for proximal tubule in acetazolamide-infused rats, the following value is obtained:

$$(H_2CO_3) = \frac{46 \ \mu Eq/min \cdot kg}{1 \times 10^{-3} \ liter/kg \cdot 3,000 \ min^{-1}} = 15.3 \ \mu Eq/liter,$$

where a tubular volume of 1 ml/kg rat is assumed. This value is somewhat lower than that published previously (Malnic and Mello Aires, 1971) because of the use of a more recently obtained and higher rate coefficient.

Using a similar calculation performed via a different method (Rector, 1973), a luminal carbonic acid concentration of the order of 20 μEq/liter would be obtained. This value should be compared to the actual luminal carbonic acid concentration of 25 μEq/liter, calculated on the basis of luminal pH and bicarbonate concentration obtained in proximal tubule of acetazolamide-infused rats (Vieira and Malnic, 1968) and using the pK of carbonic acid given by Gibbons and Edsall (1963) of 3.57. The concentration calculated by means of Eq. 6 depends not only on experimental error, but also on a rather rough estimate of total tubular volume. However, the fact that the estimated tubular acid concentration is not higher than the measured one suggests that most if not all tubular bicarbonate could be reabsorbed via H ion secretion. Since plasma concentration of carbonic acid in acetazolamide-infused rats is of the order of 5.5 μEq/liter, the existence and origin of a disequilibrium pH under these conditions is demonstrated. As stated previously, a disequilibrium pCO_2 could participate in this pH, and would likewise lead to an increase in luminal H_2CO_3 concentration. In the distal tubule of rats, by a similar method, a luminal carbonic acid concentration of 65.5 μEq/liter can be estimated on the basis of the prevailing rate of bicarbonate reabsorption, whereas the measured value is 48 μEq/liter. In view of the errors and assumptions involved, those values can be considered of the same order of magnitude, indicating that in the distal tubule H ion secretion may also account for a considerable proportion of bicarbonate reabsorption.

Analysis of Table 2 indicated that after inhibition of carbonic anhydrase a sizable rate of tubular acidification is still encountered. In the preceding paragraphs, we have shown that this residual acidification is probably mediated by H ion secretion. So it is of interest to compare that rate with the possible H ion generation via uncatalyzed CO_2 hydration expected under those circumstances. For this purpose, the rate of CO_2 hydration can be calculated by Eq. 3; using the described rate coefficient and blood CO_2 levels in acetazolamide-infused rats, we obtain the following value:

$$\dot{V}_{CO_2} = 0.15 \cdot 1.72 = 0.26 \text{ mEq/liter} \cdot \text{sec}$$

Considering that H ions for secretion are generated within the tubular cells and that according to tubular geometry (Rector, 1973) cell volume is 1.78 times greater than tubular luminal volume, we can expect a total rate of H ion generation of 0.46 mEq/liter $\cdot$ sec by this mechanism. The rate should be compared to the column in Table 2 in which H ion flow is given in mEq/liter $\cdot$ sec; it is clearly noted that apparently uncatalyzed hydration of CO_2 can account for only a minor fraction of the observed acidification rates. It could be argued that during perfusion experiments a net secretory rate of the order of the maximal flow rates given in Table 2 is only transient, the cells being able to maintain such rates for short periods of time without possessing a steady-state rate of H ion generation of that order. However, when proximal and distal reabsorption rates in free flow are calculated, using the same rate coefficients and measured mean bicarbonate concentrations, the following values are obtained:

$$\dot{V}_{HCO_3} = \frac{\ln 2}{t/2} \cdot (HCO_{3\,ff}^- - HCO_{3\,ss}^-)$$

$$= \frac{0.694}{8.12} \cdot (26.6 - 7.1) = 1.67 \text{ mEq/liter·sec}$$

where the subscripts *ff* and *ss* refer to free-flow and steady-state concentrations. A similar calculation yields a value of 2.0 mEq/liter·sec for distal tubule. Both these values are considerably higher than the uncatalyzed hydration rate.

The foregoing considerations seem to indicate that acidification in renal tubules cannot be sustained by uncatalyzed CO_2 hydration in tubular cells alone. What other mechanisms of acidification or of H ion generation could be invoked to explain the remainder of the measured acidification rates? Among several hypothetical possibilities, we will analyze the following.

1. Incomplete inhibition of carbonic anhydrase. It is generally considered that enzyme inhibition is essentially complete at the doses used, since further raising of drug levels does not lead to further inhibition of acidification (Maren, 1967). As discussed previously, Maren and Ellison (1967) have found different isoenzymes of carbonic anhydrase in supernatant (cytoplasm) and particulate (microsomal) fractions of dog and rat kidney; the microsomal enzyme has a lower sensitivity to drug inhibition than the cytoplasmic enzyme. Despite the fact that higher drug levels are necessary to produce comparable inhibition in the microsomal enzyme, almost total inhibition can still be obtained. The question remains, however, whether *in vivo* membrane-bound enzyme might not exist in a geometrical configuration that might render complete inhibition more difficult.

2. Peritubular H. The source of H ions secreted may not be limited to the tubular cell but may encompass the peritubular interstitium and peritubular capillaries. We have recently shown that tubular acidification is rapidly abolished upon perfusion of peritubular capillaries with alkaline solutions (Malnic and Giebisch, 1972*a*), suggesting rapid equilibration between cell and capillaries; the latter could represent an additional volume available for H ion generation. The magnitude of this volume is, however, difficult to assess. More direct studies on peritubular capillaries will be necessary to investigate this possibility.

3. Bicarbonate reabsorption per se. The existence of an anion-transport mechanism for bicarbonate has been postulated by Maren (1967, 1969) for kidney and by Schilb and Brodsky (1972) for turtle bladder in order to explain bicarbonate reabsorption after carbonic anhydrase inhibition. As discussed previously, the occurrence after inhibition of a disequilibrium pH as well as the sensitivity of bicarbonate reabsorption after inhibition to changes in pCO_2 (Rector et al., 1960) suggests that even after inhibition H ion secretion appears to be the most important acidification mechanism. However, the participation of bicarbonate reabsorption *per se* in a fraction of tubular reabsorption of this anion cannot be easily discarded.

4. Carbonic acid recirculation. Recently Rector (1973) proposed a mechanism by which the bicarbonate/CO_2 system could participate in tubular acidification without the need for carbonic anhydrase. After inhibition of the enzyme, as indicated above, luminal carbonic acid concentrations increase markedly. Carbonic acid could then diffuse back into the tubular cell by non-ionic diffusion along the concentration gradient; in the cell it could either dissociate — furnishing H ions for secretion — or neutralize OH^- ions formed by a proton pump splitting water; both these reactions are very fast and independent of carbonic anhydrase catalysis. More direct experimental evidence is needed, however, to demonstrate the existence of such a process.

ACKNOWLEDGMENTS

This work was supported by the São Paulo State Research Foundation (FAPESP).

REFERENCES

Bank, N., and Aynedijian, H. S. (1967): A microperfusion study of bicarbonate accumulation in the proximal tubule of the rat kidney. *J. Clin. Invest.* 46:95–102.

Beck, L. H., and Goldberg, M. (1973): Effects of acetazolamide and parathyroidectomy on renal transport of sodium, calcium and phosphate. *Am. J. Physiol.* 224:1136–1142.

Berliner, R. W., Kennedy, T. J., and Orloff, J. (1951): Relationship between acidification of the urine and potassium metabolism. *Am. J. Med.* 11:274–282.

Boulpaep, E. L., and Seely, J. F. (1971): Electrophysiology of proximal and distal tubules in the autoperfused dog kidney. *Am. J. Physiol.* 221:1084–1096.

Cassola, A. C. (1974): Avaliação da permeabilidade do epitélio de nefrons corticais de rato a ions hidrogênio e fosfato. *Thesis,* São Paulo, 1974.

Clapp, J. R., Watson, J. F., and Berliner, R. W. (1963): Effect of carbonic anhydrase inhibition on proximal tubular bicarbonate reabsorption. *Am. J. Physiol.* 205:693–696.

Durbin, R. P., and Heinz, E. (1958): Electromotive chloride transport and gastric secretion in the frog. *J. Gen. Physiol.* 41:1035–1047.

Froemter, E., Rumrich, G., and Ullrich, K. J. (1973): Phenomenologic description of Na^+, Cl^- and HCO_3^- absorption from proximal tubules of the rat kidney. *Pfluegers Arch.* 343:189–220.

Garg, L. C., and Maren, T. H. (1972): The rates of hydration of carbon dioxide and dehydration of carbonic acid at 37°C. *Biochim. Biophys. Acta* 261:70–76.

Gertz, K. H. (1963): Transtubulaere Natriumchloridfluesse und Permeabilitaet fuer Nichtelektrolyte im proximalen und distalen Konvolut der Rattenniere. *Pfluegers Arch.* 276:336–356.

Gibbons, B. H., and Edsall, J. T. (1963): Rate of hydration of carbon dioxide and dehydration at 25°C. *J. Biol. Chem.* 238:3501–3507.

Gonzalez, C. F. (1969): Inhibitory effect of acetazolamide on the active chloride and bicarbonate transport mechanisms across short-circuited turtle bladders. *Biochim. Biophys. Acta* 193:146–158.

Hansson, H. P. J. (1968): Histochemical demonstration of carbonic anhydrase activity in some epithelia noted for active transport. *Acta Physiol. Scand.* 73:427–434.

Hegel, U., Froemter, E., and Wick, T. (1967): Der elektrische Wandwiderstand des proximalen Konvolutes der Rattenniere. *Pfluegers Arch.* 294:274–290.

Karlmark, B. (1973): Intratubular carbon dioxide tension in the proximal tubule of the kidney. *Acta Physiol. Scand.* 87:54A–55A.

Karlmark, B., and Sohtell, M. (1973): The determination of bicarbonate in nanoliter samples. *Anal. Biochem.* 53:1–11.

Kitahara, S., Fox, K. R., and Hogben, C. A. M. (1967): Depression of chloride transport by carbonic anhydrase inhibitors in the absence of carbonic anhydrase. *Nature* 214:836–837.

Kunau, R. T. (1972): The influence of the carbonic anhydrase inhibitor, benzolamide (CL-11, 366), on the reabsorption of chloride, sodium and bicarbonate in the proximal tubule of the rat. *J. Clin. Invest.* 51:294–306.

Loennerholm, G. (1973): Histochemical demonstration of carbonic anhydrase activity in the human kidney. *Acta Physiol. Scand.* 88:455–468.

Loennerholm, G. (1974): Carbonic anhydrase in the cornea. *Acta Physiol. Scand.* 90:143–152.

Magid, E., and Turbeck, B. O. (1968): The rates of the spontaneous hydration of CO_2 and the reciprocal reaction in neutral aqueous solutions between 0° and 38°. *Biochim. Biophys. Acta* 165:515–524.

Malnic, G., and Giebisch, G. (1972a): Mechanism of renal hydrogen ion secretion. *Kidney Int.* 1:280–296.

Malnic, G., and Giebisch, G. (1972b): Some electrical properties of distal tubular epithelium in the rat. *Amer. J. Physiol.* 223:797–808.

Malnic, G., and Mello Aires, M. (1970): Microperfusion study of anion transfer in proximal tubules of rat kidney. *Amer. J. Physiol.* 218:27–32.

Malnic, G., and Mello Aires, M. (1971): Kinetic study of bicarbonate reabsorption in proximal tubule of the rat. *Amer. J. Physiol.* 220:1759–1767.

Malnic, G., Mello Aires, M., and Cassola, A. C. (1974a): Kinetic analysis of renal tubular acidification by antimony microelectrodes. In: *Ion-Selective Microelectrodes,* edited by N. C. Hebert and H. J. Berman, pp. 89–108. Plenum Press, New York.

Malnic, G., Mello Aires, M., de Mello, G. B., and Giebisch, G. (1975): Kinetics of H ion secretion in cortical tubules of rat kidney. *Submitted for publication.*

Malnic, G., Mello Aires, M., and Giebisch, G. (1972): Micropuncture study of renal tubular hydrogen ion transport in the rat. *Am. J. Physiol.* 222:147–158.

Malnic, G., Mello Aires, M., and Vieira, F. L. (1970): Chloride excretion in nephrons of rat kidney during alterations of acid-base equilibrium. *Am. J. Physiol.* 218:20–26.

Malnic, G., and Vieira, F. L. (1972): The antimony microelectrode in kidney micropuncture. *Yale J. Biol. Med.* 45:356–367.

Maren, T. H. (1967): Carbonic anhydrase: chemistry, physiology, and inhibition. *Physiol. Rev.* 47:595–781.

Maren, T. H. (1969): Renal carbonic anhydrase and the pharmacology of sulfonamide inhibitors. In: *Handbuch der experimentellen Pharmakologie,* edited by H. Herken, Vol. XXIV, pp. 195–256. Springer-Verlag, Berlin.

Maren, T. H., and Broder, L. E. (1970): The role of carbonic anhydrase in anion secretion into cerebrospinal fluid. *J. Pharmacol. Exp. Thera.* 172:197–202.

Maren, T. H., and Ellison, A. C. (1967): A study of renal carbonic anhydrase. *Mol. Pharmacol.* 3:503–508.

Maren, T. H., Wadsworth, B. C., Yale, E. K., and Alonso, L. G. (1954): Carbonic anhydrase inhibition. III. Effects of Diamox on electrolyte metabolism. *Bull. Johns Hopkins Hosp.* 95:277–321.

Mello Aires, M., and Malnic, G. (1972): Micropuncture study of acidification during hypochloremic alkalosis in the rat. *Pfluegers Arch.* 331:13–24.

Morel, F., and Murayama, Y. (1970): Simultaneous measurement of unidirectional and net sodium fluxes in microperfused rat proximal tubules. *Pfluegers Arch.* 320:1–23.

Rector, F. C. (1973): Acidification of the urine. In: *Handbook of Physiology,* Sect. 8, edited by J. Orloff and R. W. Berliner, pp. 431–454. American Physiological Society, Washington, D.C.

Rector, F. C., Carter, N. W., and Seldin, D. W. (1965): The mechanism of bicarbonate reabsorption in the proximal and distal tubules of the kidney. *J. Clin. Invest.* 44:278–290.

Rector, F. C., Seldin, D. W., Roberts, A. D., and Smith, J. S. (1960): The role of plasma CO_2

tension and carbonic anhydrase activity in the renal reabsorption of bicarbonate. *J. Clin. Invest.* 39:1706–1721.

Schilb, T. P., and Brodsky, W. A. (1972): CO_2 gradients and acidification by transport of HCO_3 in turtle bladders. *Am. J. Physiol.* 222:272–281.

Steinmetz, P. R., and Lawson, L. R. (1971): Effect of luminal pH on ion permeability and flows of Na^+ and H^+ in turtle bladder. *Am. J. Physiol.* 220:1573–1580.

Vieira, F. L., and Malnic, G. (1968): Hydrogen ion secretion by rat renal cortical tubules as studied by an antimony microelectrode. *Am. J. Physiol.* 214:710–718.

Ward, R. L. (1969): [35]Cl Nuclear magnetic resonance studies of a zinc metallo-enzyme, carbonic anhydrase. *Biochemistry* 8:1879–1883.

Whittembury, G., Rawlins, F. A., and Boulpaep, E. L. (1973): Paracellular pathway in kidney tubules: electrophysiological and morphological evidence. In: *Transport Mechanisms in Epithelia,* edited by H. H. Ussing and N. H. Thorn, pp. 577–588. Munksgaard, Copenhagen.

Concepts of Membranes in Regulation and Excitation,
edited by M. Rocha e Silva and G. Suarez-Kurtz.
Raven Press, New York © 1975.

Studies on the Process of Acidification in the Urinary Bladder of the Toad *In Vitro*

M. R. Furtado and A. C. Nero

Department of Internal Medicine, Faculty of Medicine of Ribeirão Preto, University of São Paulo, 14100 Ribeirão Preto, S.P., Brazil

INTRODUCTION

Urinary bladders of toads and turtles are known to be potent acidifiers (Schilb and Brodsky, 1966; Steinmetz, 1967; Frazier and Vanatta, 1971). The carbonic anhydrase activity required to catalyze CO_2 hydration for the generation of hydrogen ions has been shown to occur in the mucosal cells of turtle and toad bladders (Rosen, 1970; Schwartz, Rosen, and Steinmetz, 1972). Acidification of the mucosal fluid associated with alkalinization of the serosal solution has been observed even when CO_2 and HCO_3^- are lacking, as well as in the absence of a favorable transepithelial concentration gradient. Studies in both turtle and toad bladder seem to indicate that the movement of hydrogen ions cannot be accounted for by electrical forces only and suggest that hydrogen ions are actively transported across the bladder (Steinmetz, 1967; Frazier and Vanatta, 1971, 1972). These tissues behave similarly to the mammalian distal nephron with respect to acidifying characteristics. Moreover, the anuran urinary bladder is the phylogenetic and functional analogue of the collecting tubule of the mammalian kidney, and its structural simplicity and ease in handling make it a widely used preparation.

The studies reported here were carried out initially in the sac preparation of the isolated bladder as originally described by Bentley (1958). Later, a new chamber was designed to permit infusion of a perfused tubule-like preparation of toad bladder (to be described in detail elsewhere, Furtado, 1974).

MATERIALS AND METHODS

Bentley's Preparation

Isolated urinary bladders of *Bufo marinus paracnemis* Lutz were mounted as the sac preparation (Fig. 1), suspended in a 10-ml constant-temperature bath of amphibian phosphate-Ringer's solution (free of bicarbonate) and

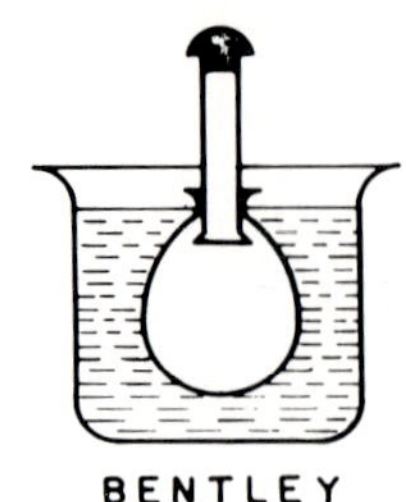

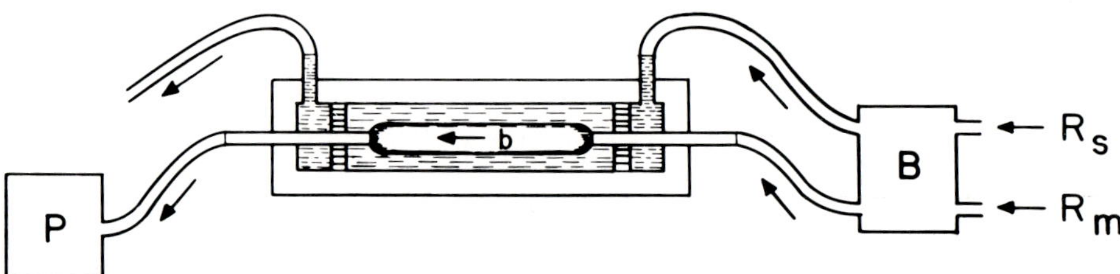

FIG. 1. *Top:* Original Bentley preparation. The bladder sac is tied to one end of a plastic cannula, the other extremity being sealed by a wax plug. Not shown are the air inlet and a hook holding the bladder sac-cannula assembly. *Bottom:* Tubular preparation of the bladder (b) inside the Lucite perfusion chamber. Roller-pump (B) provides inflowing Ringer's solutions for both infusion (R_m) and perfusion (R_s). P is the electrode-pH meter-recorder assembly.

filled with 5 ml of identical solution. The phosphate concentration was 1.5 mM. Whenever a substitute Ringer's solution was used, care was taken to maintain the same total osmolarity. The stoppered bladder sacs formed a closed system while the serosal bath was in contact with air and was stirred continuously by bubbling oxygen. Measurements of pH were made by means of a combination electrode with experimental periods of 30-min duration. Paired hemibladders were used throughout. Using the Henderson-Hasselbalch equation and the molar concentration of phosphate buffer, the amount of hydrogen added to the mucosal medium could be calculated. The results are given in μmoles of H^+/100 mg dry weight of bladder/min to express the acidifying response.

Tubular Preparation of the Bladder

In the infusion-perfusion experiments, a toad half-bladder was mounted as a tubular segment tied to the extremities of two plastic cannulas held in place within the perfusion chamber (Fig. 1). A multichannel roller-pump provided oxygenated inflowing solutions for both infusion and perfusion. The outflowing serosal perfusate could be collected for study or discarded.

The mucosal infusion fluid, after leaving the bladder tubule, is driven to a Beckman-micro blood pH assembly connected to a Beckman-Century SS-1 pH meter. The complete set-up closely approximates the conformation and working conditions of the mammalian collecting tubule and thus facilitates comparative studies. It has the additional advantage of eliminating the formation of unstirred layers of fluid adjacent to the membrane by continuously renewing the inflowing solutions and thus presenting a constant pH at the entrances to the chamber. The set-up also allows for kinetic studies and for study of the influence of solution flow rate on the acidification process.

RESULTS

By decreasing the velocity of infusion from 0.2 to 0.1 ml/min and then to 0.05 ml/min in the tubular preparation of bladder (Furtado, *submitted*), a proportionally increased acidification is obtained, owing, at least in part, to a greater contact time. The results presented below were observed with a flow rate of 0.1 ml/min and only mucosal pH changes are reported.

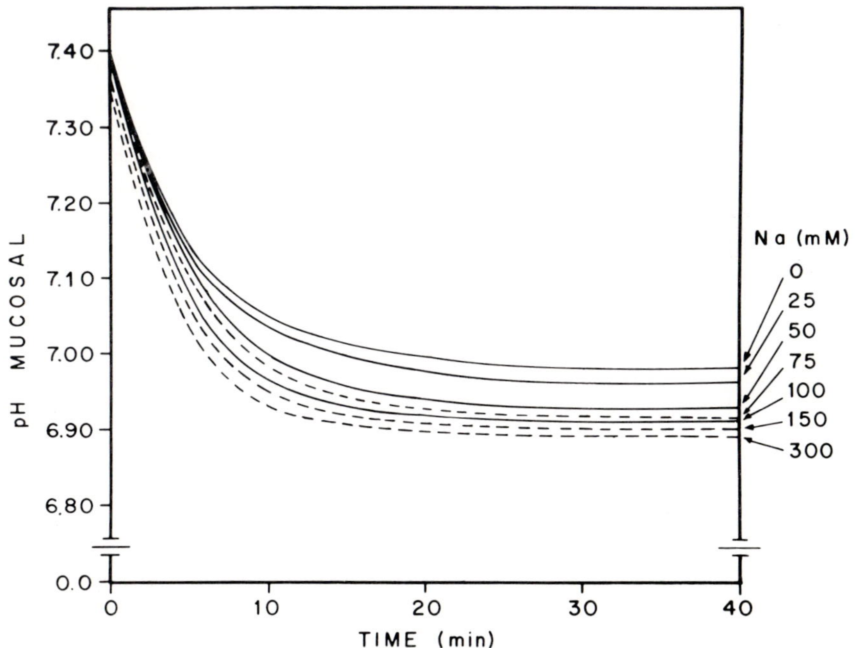

FIG. 2. Mucosal pH changes observed in tubular preparation of the bladder in the absence and in the presence of various amounts of sodium. The experiment was performed using, in sequence, two paired half-bladders (*solid and broken lines*). Note that the initial pH for the concentrations of 150 and 300 mM Na+ was slightly lower than other initial pHs. Otherwise the final steady-state pH lines did not differ from the curves obtained with 75 and 100 mM Na+ solutions.

Acidification of the mucosal fluid was slightly but consistently affected by replacement of sodium by choline chloride in the mucosal solution. Figure 2 is representative of this series of experiments, using the infusion-perfusion chamber. Starting with a sodium-free (choline-Ringer's) period, for every subsequent period an increment in sodium concentration was added to the mucosal infusion (keeping the osmolarity constant). It was observed that even in the absence of sodium, acidification proceeded at high levels, confirming previous observations (Ludens and Fanestil, 1972). Nevertheless, the acidifying response increased by elevating sodium concentration up to an inflection point around 70 mM Na^+. Above this level, even with 350 mM Na^+ (meaning, therefore, increased osmolarity), acidification did not change. In conclusion, apparently only a minor portion of acidification seems to be coupled to sodium transport.

Substances known to inhibit sodium reabsorption, therefore, would probably decrease acidification. Several natriuretic agents have been shown to block the mucosal-to-serosal sodium flux across the skin and the bladder of anurans (Pendleton, Sullivan, and Tucker, 1968; Gatzy, 1971; Ludens and Fanestil, 1972). Experiments were performed with acetazolamide, a potent

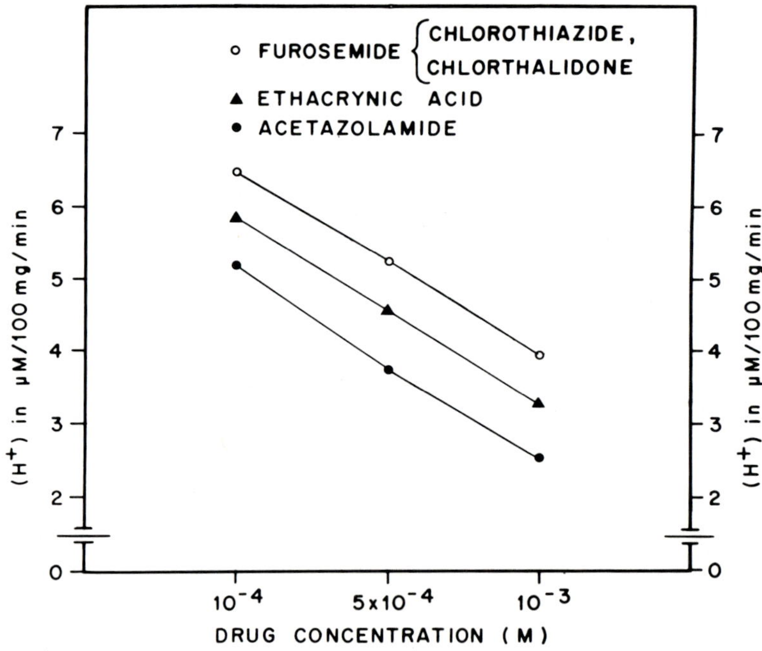

FIG. 3. Dose-response relationships for the diuretics and their effects on acid formation. Each point is the mean of 10 determinations using Bentley's preparation of bladders. Symbols as identified in the figure. Results obtained with chlorothiazide and chlorthalidone were about the same as those observed with furosemide and yielded identical dose-effect relationships.

carbonic anhydrase inhibitor; furosemide, chlorthalidone, and chlorothiazide, which are weaker inhibitors of this enzyme; and ethacrynic acid, an agent devoid of carbonic anhydrase-inhibiting activity. In the sac preparation of toad bladder, when applied to the serosal fluid in the concentrations of 10^{-4}, 5×10^{-4}, and 10^{-3} M, these drugs inhibited differently the rate of acidification of the mucosal solution (Fig. 3). Acetazolamide was by far the most potent inhibitor, probably on account of its greater activity as a carbonic anhydrase inhibitor. Figures 4 and 5 show the typical responses to acetazolamide and to furosemide, respectively, in infusion-perfusion experiments when the drugs were applied either to the serosal or to the mucosal solutions. The striking feature was the conspicuous inhibition observed when these agents were in contact with the mucosal epithelium of the membrane.

Spironolactone (aldactone) was also tested, in sac preparations, in the serosal as well as in the mucosal fluid and did not affect the rate of acidification by the bladder.

Replacement of chlorides on both sides of the bladder by sulfate salts, using the sac preparation, had a slight but consistent inhibitory effect on the acidifying activity of the bladder. In the presence of chloride, the acid flux was 12.05 μM/100 mg tissue/min, whereas in the absence of chloride it was only 9.06 μM/100 mg/min in simultaneous and paired experiments. Although perhaps equivocal, these results prompted another series of experiments with acetazolamide and ethacrynic acid, utilizing the sac preparation and

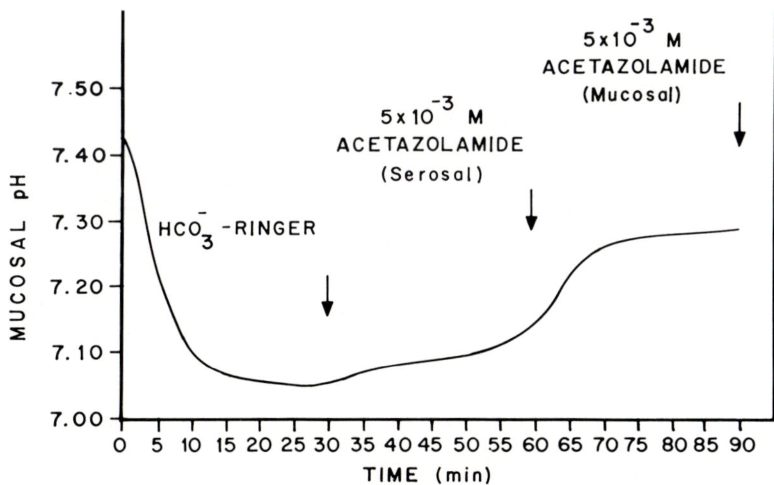

FIG. 4. Tracing of observed pH changes in the mucosal solution in the tubular preparation of the bladder. Typical experiment using bicarbonate-Ringer's solution and showing the influence of 5×10^{-3} M acetazolamide (*intervals between arrows*) when applied to the serosal or to the mucosal fluids. Solutions containing the drug had their pH adjusted to that of pure Ringer's solution.

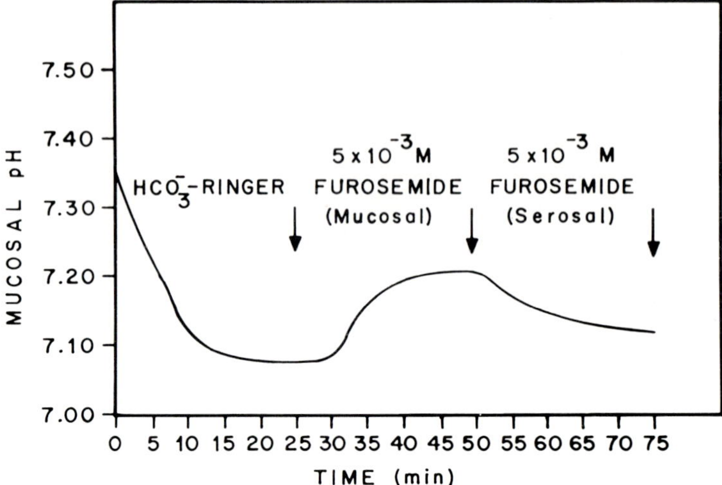

FIG. 5. Tracing of a typical infusion-perfusion experiment showing the inhibition of acidification by furosemide when applied to mucosal or serosal Ringer's solution. For details see Fig. 4.

applying the drugs only to the serosal side of the bladder in the presence (Cl-Ringer's) and in the absence (SO$_4$-Ringer's) of chlorides, showing a different behavior of these two drugs, as depicted in Fig. 6. Acetazolamide was more efficient as an inhibitor in the presence of chloride. On the other hand, the inhibiting activity of ethacrynic acid, when applied to the serosal side of the bladder, was more pronounced in the absence of chloride.

As a whole, these results emphasize the importance of chloride ions in the mechanism of acidification and suggest that, at least in part, the inhibitory effect of acetazolamide seems to be dependent on the presence of chloride ions. The latter observation corroborates data indicating that acetazolamide and carbonic anhydrase have their own effect on the movement of chloride ions across the membrane, possibly coupled to the transport of hydrogen ions in the secretory direction (Kashgarian, Warren, and Levitin, 1965; Hogben, 1967; Malnic and Mello Aires, 1970). The reason ethacrynic acid was more active in the absence of chloride cannot be explained at present.

DISCUSSION

As shown above, acetazolamide, like furosemide, becomes a more potent *inhibitor* of acidification when added to the mucosal side of the bladder. Ethacrynic acid, on the other hand, when applied to the mucosal solution is an efficient *activator* of acidification, as shown in Fig. 7, which depicts a typical experiment with the infusion-perfusion chamber. It seems that ethacrynic acid when present in the serosal fluid inhibits acidification, possibly

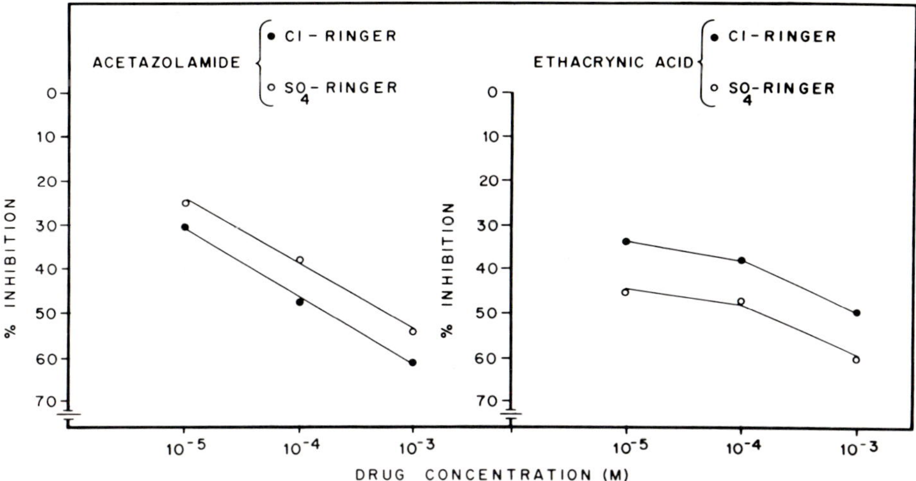

FIG. 6. Dose-response relationships for acetazolamide and ethacrynic acid and their effects on the acidification process as judged from the inhibition observed when applied to the serosal fluid at various concentrations. Both agents were tested in the presence (Cl-Ringer's) and absence (SO$_4$–Ringer's) of chlorides, using the bladder-sac preparation. The % inhibition was calculated for each experimental group taking the respective paired-control period as having 100% activity. Each point is the mean of 5 experiments.

by a blocking action on sodium reabsorption (which is small in magnitude). When in contact with the mucosal side of the bladder, ethacrynic acid stimulates the hydrogen ion secretory mechanism, which could be coupled to chloride secretion, and is located, presumably, in the external membrane of the bladder cell.

It is well known that administration in man of either ethacrynic acid or furosemide may lead to alkalosis. A "contraction" alkalosis was first advanced to explain this observation (Cannon, Heinemann, Albert, Laragh, and Winters, 1965). An augmented supply of sodium to the Na^+-H^+-K^+ exchange site in the distal tubule of the kidney (caused by the blockade on sodium reabsorption proximal to the exchange site), leading to increased urinary acidification, is common to both diuretic agents (Puschett and Goldberg, 1968; Stein, Wilson, and Kirkendall, 1968). In the case of ethacrynic acid, the latter effect would enhance the intrinsic acidifying properties of the drug, with a resultant tendency to deplete hydrogen ion stores of the body.

It was observed that furosemide inhibited the process of acidification in the toad bladder, and it could also do so in the mammalian distal nephron. However, that effect could be overcome in terms of magnitude by furosemide's action on sodium reabsorption proximal to the exchange site (see above) and by its phosphaturic action (inhibition of the proximal reabsorption of phosphate, as demonstrated by Puschett and Goldberg, 1968).

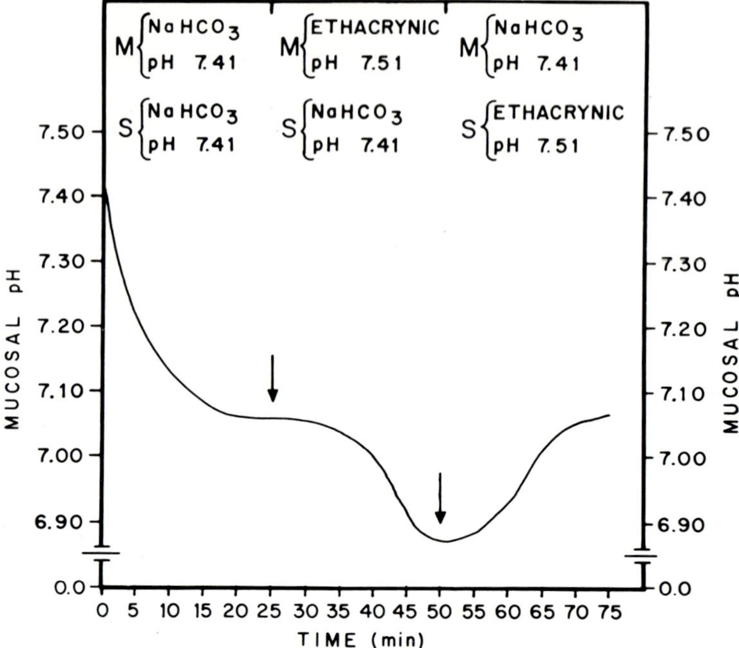

FIG. 7. Tracing of mucosal pH changes taken from a representative experiment using the tubular preparation of the bladder and showing the effect of ethacrynic acid when applied to either side of the membrane. Note that ethacrynic acid when used in the mucosal fluid provoked a further decrease in mucosal pH. Doses: ethacrynic acid 5×10^{-3} M.

Therefore, the increased delivery of phosphate to the distal nephron, acting as both impermeant anion (thus increasing intratubular negativity) and buffer, would stimulate hydrogen ion secretion (Bank and Schwartz, 1960; Malnic and Giebisch, 1972).

ACKNOWLEDGMENT

This work has been supported by a research grant from the São Paulo State Research Foundation. (FAPESP).

REFERENCES

Bank, N., and Schwartz, W. B. (1960): The influence of anion penetrating ability on urinary acidification and the excretion of titratable acid. *J. Clin. Invest.* 39:1516–1525.

Bentley, P. J. (1958): The effects of neurohypophyseal extracts on water transfer across the wall of the isolated urinary bladder of the toad *Bufo marinus. J. Endocrinol.* 17:201–209.

Cannon, P. J., Heinemann, H. O., Albert, M. S., Laragh, J. H., and Winters, R. W. (1965): "Contraction" alkalosis after diuresis of edematous patients with ethacrynic acid. *Ann. Intern. Med.* 62:979–990.

Frazier, L. W., and Vanatta, J. C. (1971): Excretion of H⁺ and NH₄⁺ by the urinary bladder of the acidotic toad and the effect of short-circuit current on the excretion. *Biochim. Biophys. Acta* 241:20–29.

Frazier, L. W., and Vanatta, J. C. (1972): Mechanism of acidification of the mucosal fluid by the toad urinary bladder. *Biochim. Biophys. Acta* 290:168–177.

Furtado, M. R.: An infusion-perfusion chamber to accommodate a tubular preparation of toad bladder. *Submitted for publication.*

Gatzy, J. T. (1971): The effect of K^+-sparing diuretics on ion transport across the excised toad bladder. *J. Pharmacol. Exp. Ther.* 176:580–594.

Hogben, C. A. M. (1967): The chloride effect of carbonic anhydrase inhibitors. *Mol. Pharmacol.* 3:318–326.

Kashgarian, M., Warren, Y., and Levitin, H. (1965): Micropuncture study of proximal renal tubular chloride transport during hypercapnea in the rat. *Am. J. Physiol.* 209:655–658.

Ludens, J. H., and Fanestil, D. D. (1972): Acidification of urine by the isolated urinary bladder of the toad. *Am. J. Physiol.* 223:1338–1344.

Malnic, G., and Giebisch, G. (1972): Mechanism of renal hydrogen ion secretion. *Kidney Int.* 1:280–296.

Malnic, G., and Mello Aires, M. (1970): Microperfusion study of anion transfer in proximal tubules of rat kidney. *Am. J. Physiol.* 218:27–32.

Pendleton, R. G., Sullivan, L. P., and Tucker, J. M. (1968): The effect of chlormerodrin on the isolated toad bladder. *J. Pharmacol. Exp. Ther.* 164:362–370.

Puschett, J. B., and Goldberg, M. (1968): The acute effects of furosemide on acid and electrolyte excretion in man. *J. Lab. Clin. Med.* 71:666–677.

Rosen, S. (1970): Localization of carbonic anhydrase activity in transporting urinary epithelia. *J. Histochem. Cytochem.* 18:668–670.

Schilb, T. P., and Brodsky, W. A. (1966): Acidification of mucosal fluid by transport of bicarbonate ion in turtle bladder. *Am. J. Physiol.* 210:997–1008.

Schwartz, J. H., Rosen, S., and Steinmetz, P. R. (1972): Carbonic anhydrase function and the epithelial organization of H^+ secretion in turtle urinary bladder. *J. Clin. Invest.* 51:2653–2662.

Stein, J. H., Wilson, C. B., and Kirkendall, W. M. (1968): Differences in the acute effects of furosemide and ethacrynic acid in man. *J. Lab. Clin. Med.* 71:654–665.

Steinmetz, P. R. (1967): Characteristics of hydrogen ion transport in urinary bladder of water turtle. *J. Clin. Invest.* 46:1531–1540.

Intercellular Communication in Lymphocytes

Gilberto M. de Oliveira-Castro, Marcello A. Barcinski, and
Ionice F. Gaziri

Institute of Biophysics, Federal University of Rio de Janeiro, Rio de Janeiro, Brazil

Intercellular communication can be achieved by means of secreted substances or through direct contact. Direct cell-to-cell communication has been found in most organized tissues (Loewenstein, 1966; Furshpan and Potter, 1968; Socolar, 1973; Bennett, 1973). In most cases a high transjunctional permeability was demonstrated by the intracellular application of electric current and measurement of the resulting potential changes in the same cell and in the adjacent one. When both cells have comparable surfaces in contact with the external bathing solution, input resistances at both sides of the junction are of the same order of magnitude and electrical measurements provide a reliable index of transjunctional permeability (Sheridan, 1973).

Most of the junctions that have been analyzed are permeable to substances of molecular weight 500 and some allow passage of molecules as large as 10,000 daltons (Kanno and Loewenstein, 1966; Loewenstein, 1973). This situation allows diffusional equilibration of small intracellular ions and it is likely that metabolites or other molecular signals permeating the junctional membrane act in controlling cellular growth and differentiation (Loewenstein, 1968; Subak-Sharpe, Burk, and Pitts, 1969; Wolpert, 1969; Crick, 1970).

Some remarkable characteristics of the direct cell-to-cell communication systems are as follows. (a) Low-resistance pathways can be formed at a very fast rate—30 sec in newt embryonic cells (Ito and Loewenstein, 1969). (b) There is no need for preexisting membrane specializations in order to establish permeable junctions because manipulating cells into contact at arbitrarily chosen spots results in intercellular communication (Loewenstein, 1967). (c) Although recognition of specific cells or cell types is one of the basic roles of plasma membrane, cell coupling can result from contact between cells of diverse types and species Michalke and Loewenstein, 1971). (d) Junctional membrane permeability falls markedly when the Ca^{++} cytoplasmic concentration is increased by intracellular injection (Loewenstein, Nakas, and Socolar, 1967) or by penetration from the bathing medium (Oliveira-Castro and Loewenstein, 1971). (e) In salivary gland cells, cooling or poisoning of the energy metabolism on which Ca^{++} removal from cytoplasm depends causes great depression of transjunctional permeability (Politoff, Socolar, and Loewenstein, 1969). This depression has recently been shown, with the aid of the calcium-sensitive luminescent protein aequorin, as an indicator, to be associated with increase in Ca^{++} concentration (Loewenstein, 1973). (f) In salivary glands, depolarizing current flow through the junctional membrane depresses the junctional conductance (Socolar and Politoff, 1971), whereas the application of hyperpolarizing current into previously uncoupled cells results in junctional recoupling (Rose, 1970). Certain malignant cells in contact with each other or with their normal counterpart, however, are not electrically coupled (Borek, Higashino, and Loewenstein, 1969).

These cells also fail to pass fluorescein (300 daltons) and nucleotides (Azarnia and Loewenstein, 1971). Recent genetic analysis in two types of noncoupling cancer cells strongly suggests that in this particular cancer type the defect in growth control and the defect in junction are genetically related (Azarnia, Larsen, and Loewenstein, 1974). This finding has major general implications in cellular growth and differentiation, and bears directly on the results with differentiating lymphocytes to be described below.

The structural basis of the high-permeability junction is not definitively established. In mammalian cells the gap junction is a good candidate to mediate cell coupling. Extensive evidence suggests this morphophysiological correlation (Gilula, Reeves, and Steinbach, 1972; McNutt and Weinstein, 1970; Pinto da Silva and Gilula, 1972) by means of genetic manipulation of a hybrid cell system, the disappearance and reappearance of gap junction (Azarnia et al., 1974).

In the last decade it has become evident that some antibody responses require interaction between two classes of lymphocytes: thymus-derived (T) lymphocytes and bone marrow-derived (B) lymphocytes (for a review see Claman and Chaperon, 1969; Raff, 1973). Some basic immunological phenomena can be explained in terms of models using cell cooperation and the great amount of experimental data produced in recent years is a consequence of the renewed interest in lymphocyte membrane physiology. The plant-derived mitogen phytohemagglutinin (PHA) stimulates lymphocytes, producing blastic transformation and intensive cell division. Although the exact mechanism of stimulation is still obscure, this membrane-mediated phenomenon is in most instances associated with cell clump formation and increase in cell density in turn enhances the mitotic activity induced by PHA stimulation (Peters, 1974).

One of the early events of lymphocyte stimulation is the formation of high-permeability junctions (Hülser and Peters, 1972; Oliveira-Castro, Barcinski, and Cukierman, 1973). In this chapter we deal with the significance of electrophysiological measurements of transjunctional permeability, the physiological mechanisms that can affect this kind of intercellular communication, and the ultrastructural aspects of the close membrane appositions.

When human lymphocytes separated from peripheral blood are cultivated in Roswell Park Memorial Institute (RPMI) medium containing 20% fetal calf serum, highly permeable junctions can be detected after 1 hr of PHA stimulation. Hülser and Peters (1972), using bovine lymphocytes, detected junctions in this system after 12 min of stimulation. With increasing time in culture (up to 65 hr), a concomitant increase in the coupling coefficient is observed (Table 1).

TABLE 1. *Coupling coefficient after different periods of stimulation*

Time in culture (hr)	Percent of coupled cell pairs	V_2/V_1[a]		Number of pairs tested
		$\overline{X} \pm SD$	Range	
1	14.3	0.18 ± 0.08	0.09–0.30	14
24	45.5	0.38 ± 0.06	0.17–0.45	11
48	75	0.42 ± 0.05	0.29–0.51	16
65	80	0.41 ± 0.07	0.30–0.56	10

[a] Coupling coefficient.

At several incubation times, the same kind of preparation used for electrophysiological determination was fixed for ultrastructural studies, according to the following protocol, modified from Hirsch and Fedorko (1968): immediately before use, 2 volumes of 2.5% glutaraldehyde in 0.1 M cacodylate buffer (pH 7.4) were added to 1 volume of 1% osmium tetroxide, at 0°C. The fixative was gently added to the lymphocytes after removal of the culture medium and replaced after 1 hr with Earl's salt solution. The cell suspension was then stained with 0.25% uranyl acetate in 0.1 M acetate buffer (pH 6.3). Constant temperature (0°C) was maintained throughout the fixation and staining procedures. After being embedded in Epon mixture, the cells were sectioned with an Ultratome III. Thin sections showing silver interference colors were stained with 0.4% lead citrate and examined with an EM 6-B electron microscope.

An increase in the area of close membrane contacts could be clearly seen as the time in culture increased from 1 to 48 hr. After 48 hr cells presented a tissuelike structure (Fig. 1).

The only type of specialized cell junction found in our preparations were gap junctions characterized by a heptalaminar structure 200 to 210 Å thick and with 20 to 40 Å between the two outer electron-dense lines. In some pictures, electron-dense particles appeared in the median line, with a regular period of 80 to 90 Å from center to center (Fig. 2).

To compare the ion-dependent control of specialized junctions in human

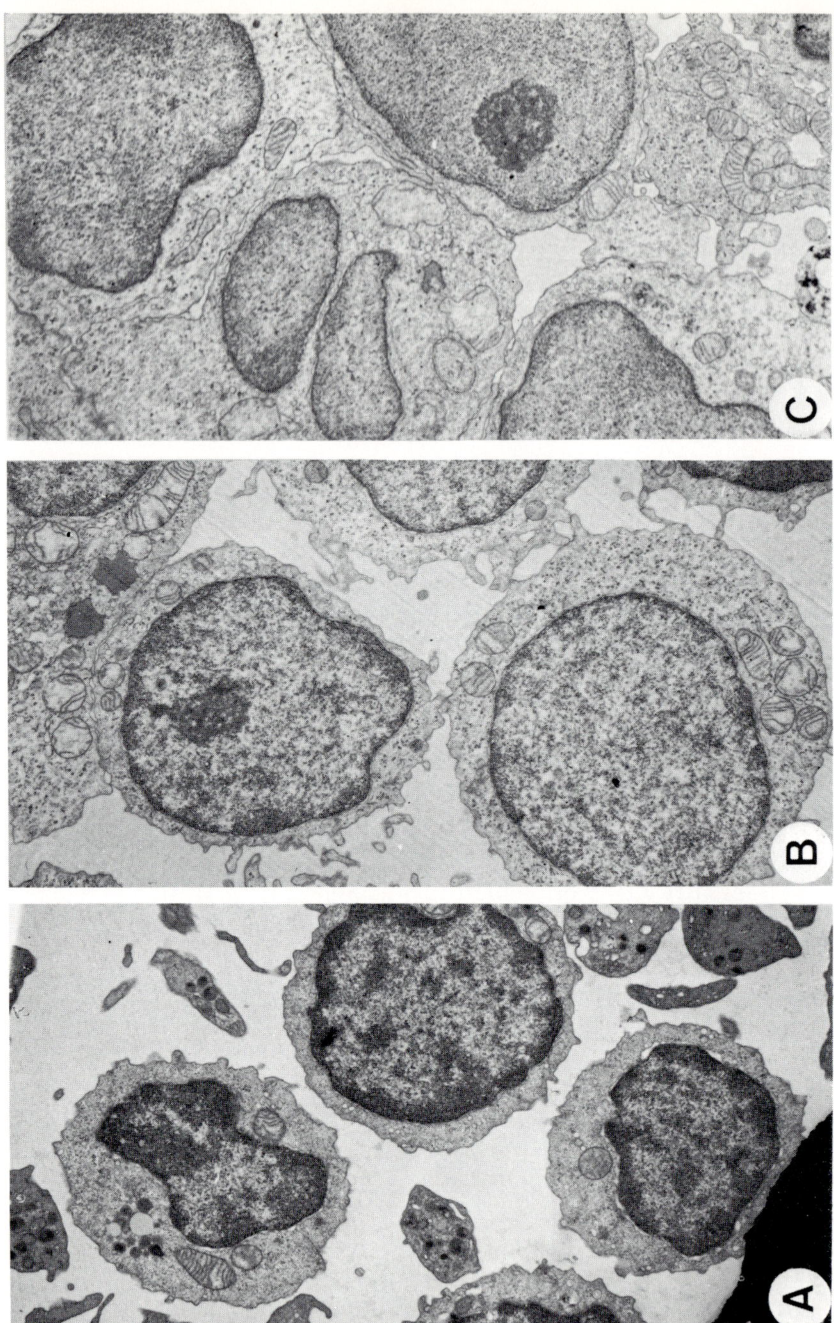

FIG. 1. Lymphocyte cultures after different periods of stimulation by PHA (×6,000). *A:* Control culture after 24 hr of incubation without PHA. Note that no cell contacts are seen. *B:* Culture after 24 hr of incubation with PHA. Several contacts between the stimulated cells are present. *C:* Culture after 65 hr of incubation with PHA. All the cells are in contact and the culture has a tissue-like appearance.

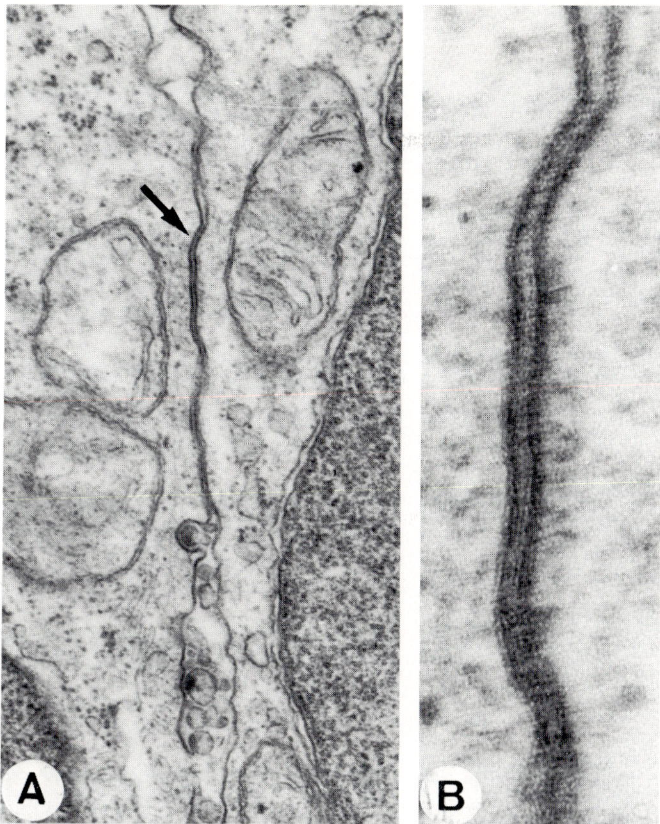

FIG. 2. Junction between stimulated lymphocytes after 65 hr of incubation with PHA. *A:* Portion of the cell junction (×25,000). *B:* Higher magnification of Fig. 2A (*arrow*). Note the appearance of the junctional complex (×200,000).

peripheral blood lymphocytes with that of the better-known epithelial cell systems, we used intracellular iontophoretic application of Ca^{++} ions. The current-voltage relationship of interlymphocyte junctions is perfectly ohmic, and the flow of current itself does not change the junctional conductance (Oliveira-Castro et al., 1973).

To test junctional permeability after a period of current injection a series of experiments with two electrodes in neighboring cells was performed. Figure 3 illustrates one of these experiments in which the cell clumps were immobilized by gentle suction (*inset*) using a larger pipette (tip diameter about 5 μm). Voltage drops (V_1, V_2) were stable throughout the experiment, with a coupling ratio of 0.56. Small differences in transmembrane potentials in both cells and the simultaneous fluctuation seen in E_1 and E_2 can be regarded as another indication of intercellular communication. Immediately

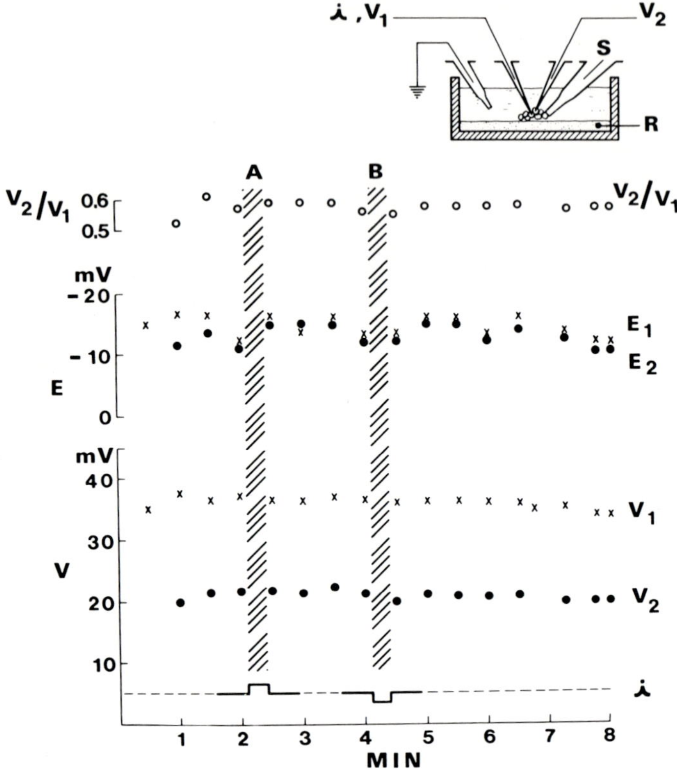

FIG. 3. Effect of current flow on junctional conductance. Transmembrane potentials (E_1, E_2) and voltage drops (V_1, V_2) resulting from current injection through KCl-filled microelectrode in Cell 1. The upper plot represents the coupling coefficient (V_2/V_1). Current pulses (*i*) applied at A (inward current) and B (outward current) did not affect intercellular conductance (pulses of 50 msec, 10^{-8} A at a rate of 5 pulses/sec during 10 sec). *Inset:* Scheme of experimental set-up showing a lymphocyte clump resting on a neutral resin (R) and immobilized by the suction pipette (S). Microelectrodes (V_1 and V_2) and reference pipette (connected to ground) are also shown. (From Oliveira-Castro and Barcinski, 1974.)

after the inward or outward current injection no changes of junctional conductance were detected.

In another group of experiments designed to test junctional permeability during current flow, electrodes were connected to a bridgelike arrangement (for details see Oliveira-Castro and Barcinski, 1974) that allows simultaneous current injection and recording. With this set-up, current pulses of 80 msec were injected into cell 1, while synchronous pulses of 20 msec were superimposed in the opposite direction through the electrode impaling cell 2. Figure 4 shows the result of one of those experiments in which junctional conductance was probed during a hyperpolarizing current pulse. Note that the coupling coefficient of approximately 0.3 applies to both long and short pulses.

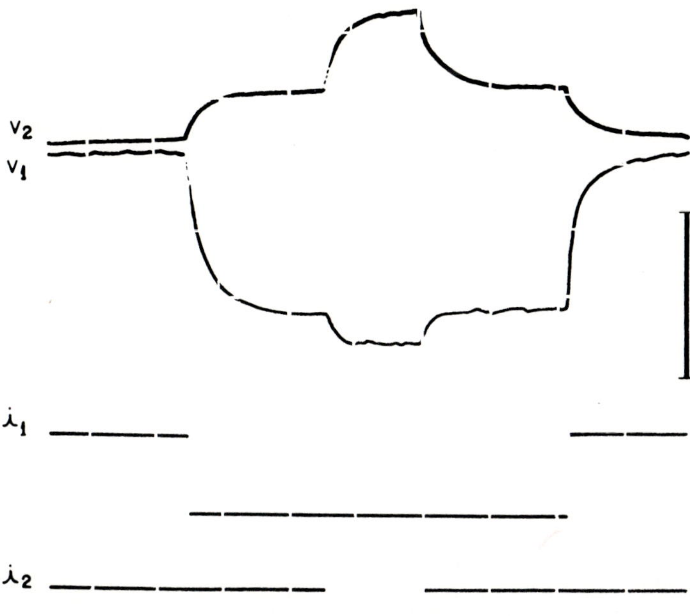

FIG. 4. Persistence of coupling ratio during inward current flow through the junction. In this experiment the current pulse i (80 msec) was injected into Cell 1 causing the voltage drops V_1 and V_2 (recorded with an inverted scale in Cell 2). During the application of the pulse i, another probing current (i_2 of 20 msec) was injected into Cell 2; this flows across the junction in the opposite direction. For both pulses the coupling ratio is about 0.3. Calibration bar 20 mV and 5×10^{-9} A. (From Oliveira-Castro and Barcinski, 1974.)

In similar experiments using depolarizing pulses no changes of transjunctional conductance were detected. These findings provide additional evidence for the view that the interlymphocyte junction behaves like an ohmic conductor.

When current is used to drive calcium ions into the cytoplasm, a striking decrease of the junctional permeability is observed after a latent period of 2 to 4 min. Figure 5 shows a progressive increase in V_1, a simultaneous decrease in V_3, and consequently a gradual reduction of the coupling coefficient to zero, indicating sealing of the communication.

Both the constant junctional conductance during and after current flow and the effect of calcium application are seen consistently in different preparations.

It has been proposed (Socolar and Politoff, 1971) that in the uncoupling procedure by outward pulses in salivary gland cells a current-induced release of Ca^{++} ions would be the mediator of the process. The next step would be the binding of the divalent cation by membrane (including its junctional portions), resulting in the decrease of transjunctional conductance. In the framework of Loewenstein's (1966) Ca hypothesis, the next step would be

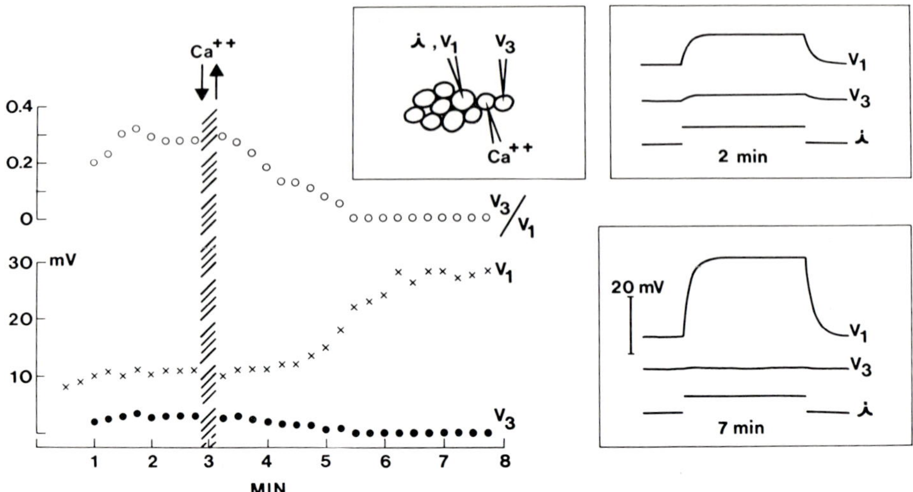

FIG. 5. Effect of calcium injection on junctional conductance. Outward pulses of 50 msec, 2.5 10^{-8} A at a rate of 10 pulses/sec during 10 sec were applied with Ca^{++}-filled electrode in Cell 2. Note progressive uncoupling with simultaneous increase of V_1 and decrease of V_3. *Insets:* Samples of oscilloscope records (retouched) of the voltage drops (V_1, V_3) and current (i) before (2 min) and after (7 min) calcium injection. Current pulse duration 50 msec and amplitude 10^{-8} A. Schematic drawing of the cell clump with position of the electrodes (see text for details). (From Oliveira-Castro and Barcinski, 1974.)

the binding of the divalent cation by membrane (including its junctional portions), resulting in the decrease of transjunctional conductance. The failure of current to induce uncoupling in the present system may mean that either current does not release calcium or perhaps there are less intracellular storage sites for this ion in the lymphocyte.

Our findings can be considered as conclusive evidence that with proper stimulation, free-floating lymphocytes are capable of establishing highly permeable transmembrane junctions as one of the first activation phenomena.

As the only junctional membrane specialization found in electrically communicating lymphocytes were gap junctions, we believe that they are responsible for the intercellular permeability.

Moreover, if PHA stimulation is a valid model for cellular interactions in immunological systems, we propose that information transfer through gap junctions may be involved in the acquisition of immunocompetence.

ACKNOWLEDGMENTS

The work reported here was supported by the Brazilian National Research Council (CNPq), the Brazilian National Bank for Economic Development (FUNTEC-241) and the Council for Graduate Education of the Federal University of Rio de Janeiro.

REFERENCES

Azarnia, R., Larsen, W. J., and Loewenstein, W. R. (1974): The membrane junctions in communicating and noncommunicating cells, their hybrids and segregants. *Proc. Nat. Acad. Sci.* 71:880–884.

Azarnia, R., and Loewenstein, W. R. (1971): Intercellular communication and tissue growth: V-A cancer cell strain that fails to make permeable membrane junctions. *J. Membr. Biol.* 6:368–385.

Bennett, M. V. L. (1973): Function of electrotonic junctions in embryonic and adult tissues. *Fed. Proc.* 32:65–75.

Borek, C., Higashino, S., and Loewenstein, W. R. (1969): Intercellular communication and tissue growth. IV. Conductance of membrane junctions of normal and cancerous cells in culture. *J. Membr. Biol.* 1:274–293.

Claman, H. N., and Chaperon, E. A. (1969): Immunologic complementation between thymus and marrow cells—a model for two-cell theory of immunocompetence. *Transplant. Rev.* 1:92–115.

Crick, F. (1970): Diffusion in embryogenesis. *Nature* 225:420–422.

Furshpan, E. J., and Potter, D. D. (1968): Low resistance junctions between cells in embryos and tissue culture. In: *Current Topics in Development Biology,* edited by A. A. Moscona and A. Monroy, Vol. 3, pp. 95–127. Academic Press, New York.

Gilula, N. B., Reeves, O. R., and Steinbach, A. (1972): Metabolic coupling, ionic coupling and cell contacts. *Nature* 235:262–265.

Hirsch, J. G., and Fedorko, M. E. (1968): Ultrastructure of human leukocytes after simultaneous fixation with glutaraldeyde and osmium tetroxide and "post fixation" in uranyl acetate. *J. Cell Biol.* 38:615–628.

Hülser, D. F., and Peters, J. H. (1972): Contact cooperation in stimulated lymphocytes: II. Electrophysiological investigations on intercellular communication. *Exp. Cell Res.* 74:319–326.

Ito, S., and Loewenstein, W. R. (1969): Ionic communication between early embryonic cells. *Dev. Biol.* 19:228–243.

Kanno, Y., and Loewenstein, W. R. (1966): Cell-to-cell passage of large molecules. *Nature* 212:629–630.

Loewenstein, W. R. (1966): Permeability of membrane junctions. *Ann. N.Y. Acad. Sci.* 137:441–472.

Loewenstein, W. R. (1967): On the genesis of cellular communication. *Dev. Biol.* 15:503–520.

Loewenstein, W. R. (1968): Some reflections on growth and differentiation. *Perspect. Biol. Med.* 11:260–272.

Loewenstein, W. R. (1973): Membrane junctions in growth and differentiation. *Fed. Proc.* 32:60–64.

Loewenstein, W. R., Nakas, N., and Socolar, S. J. (1967): Junctional membrane uncoupling: Permeability transformations at a cell membrane junction. *J. Gen. Physiol.* 50:1865–1891.

McNutt, N. S., and Weinstein, R. S. (1970): The ultrastructure of the nexus. A correlated thin section and freeze study. *J. Cell Biol.* 47:666–688.

Michalke, W., and Loewenstein, W. R. (1971): Communication between cells of different types. *Nature* 232:121–122.

Oliveira-Castro, G. M., and Barcinski, M. (1974): Calcium-induced uncoupling in communicating human lymphocytes. *Biochim. Biophys. Acta* 352:338–343.

Oliveira-Castro, G. M., Barcinski, M., and Cukierman, S. (1973): Intercellular communication in stimulated human lymphocytes. *J. Immunol.* 111:1616–1619.

Oliveira-Castro, G. M., and Loewenstein, W. R. (1971): Junctional membrane permeability: Effects of divalent cations. *J. Membr. Biol.* 5:51–77.

Peters, J. H. (1974): On the hypothesis of cell contact mediated lymphocyte stimulation. In: *Lymphocyte Recognition and Effector Mechanisms,* edited by K. Lindahl-Kiessling and D. Osoba, pp. 13–17. Academic Press, New York.

Pinto da Silva, P., and Gilula, N. B. (1972): Gap junctions in normal and transformed fibroblasts in culture. *Exp. Cell Res.* 71:393–401.

Politoff, A. L., Socolar, S. J., and Loewenstein, W. R. (1969): Permeability of a cell membrane junction: Dependence on energy metabolism. *J. Gen. Physiol.* 53:498–515.

Raff, M. C. (1973): T and B lymphocytes and immune response. *Nature* 242:19–21.

Rose, B. (1970): Junctional membrane permeability: restoration by repolarizing current. *Science* 169:607–609.

Sheridan, J. D. (1970): Electrical coupling between cancer cells in various solid tumors. *J. Cell Biol.* 45:91–99.

Sheridan, J. D. (1973): Functional evaluation of low resistance junction: influence of cell size and shape. *Am. Zool.* 13:1119–1128.

Socolar, S. J. (1973): Cell coupling in epithelia. *Exp. Eye Res.* 15:693–698.

Socolar, S. J., and Politoff, A. L. (1971): Uncoupling cell junctions in a glandular epithelium by depolarizing current. *Science* 172:492–494.

Subak-Sharpe, H., Burk, R. R., and Pitts, T. D. (1969): Metabolic cooperation between biochemically marked cells. *J. Cell Sci.* 4:353–364.

Wolpert, L. (1969): Positional information and the spatial pattern of cellular differentiation. *J. Theor. Biol.* 25:1.

Concepts of Membranes in Regulation and Excitation,
edited by M. Rocha e Silva and G. Suarez-Kurtz.
Raven Press, New York © 1975.

A Model for Angiotensin Receptors in Smooth Muscle Cells

A. C. M. Paiva and Therezinha B. Paiva

*Department of Biophysics and Physiology, Escola Paulista de Medicina,
04023 São Paulo, S.P., Brazil*

INTRODUCTION

The octapeptide hormone angiotensin II (AII) (Fig. 1) is one of the most potent myotropic substances known, producing half-maximal contraction of isolated smooth muscle at concentrations of the order of 10^{-9} M. Available evidence indicates that the first step in the chain of events leading to contraction is the interaction of the hormone with the plasma membrane of the smooth muscle cell (Rasmussen, 1972). Another step in this chain of events is probably the liberation of Ca^{++} to the environment of the contractile protein initiating contraction (Somlyo and Somlyo, 1968).

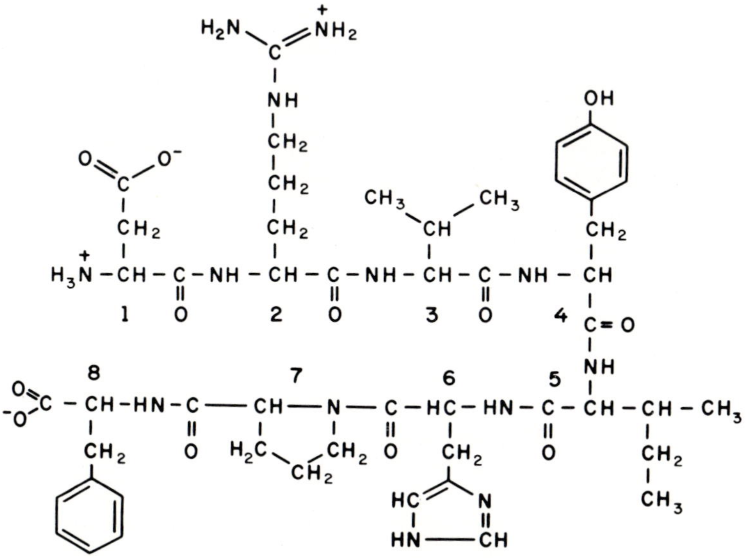

FIG. 1. Structural formula of Ile[5]-angiotensin II (AII).

ATTEMPTS TO ISOLATE AND CHARACTERIZE AII RECEPTORS

Definitive understanding of the mechanism of the myotropic action of AII awaits the isolation and characterization of its receptor on the cell membrane; and efforts in this direction are being made. Baudouin, Meyer, Fermandjian, and Morgat (1972) have isolated from rabbit aorta a microsomal preparation capable of releasing incorporated calcium upon addition of angiotensin. A membrane fraction obtained from rabbit aorta bound tritiated angiotensin with affinity constants of the same order as the AII concentration that induces half-maximal contraction of the intact artery *in vitro* (Devynck, Pernollet, Meyer, Fermandjian, and Fromageot, 1973). An angiotensin-binding component of this membrane was solubilized and postulated to contain the receptors (Devynck, Pernollet, Meyer, Fermandjian, Fromageot, and Bumpus, 1974).

An attempt to isolate angiotensin receptors from the smooth muscle of guinea pig ileum is also under way. A plasma membrane preparation was obtained that responded to AII with radial contraction (Oliveira and Holzhacker, 1974). Spin-label studies showed that addition of AII to these membranes induced conformational changes that might be among the events linking hormone-receptor interaction with muscle contraction (Schreier-Muccillo, Niculitcheff, Oliveira, Shimuta, and Paiva, 1974).

These efforts toward isolating a chemically defined species that can be recognized as part of the receptor in the membrane constitute a valid approach to the understanding of the mechanism of hormone action and should be pursued. However, the results presently available indicate that the receptor is a complex entity that may be substantially altered or even completely destroyed by the disruption of cell integrity in attempts to purify it. This is illustrated by the discrepancies between the relative affinities of several angiotensin analogues and homologues for the membranes and solubilized fractions from rabbit aorta (Devynck et al., 1974) and the affinities deduced from biological activity upon the intact tissue (Paiva, Goissis, Juliano, Miyamoto, and Paiva, 1974*a*). The conformational changes observed in spin-labeled guinea pig ileum membranes (Schreier-Muccillo et al., 1974) occurred at AII concentrations much higher than those needed in the isolated muscle bath to produce contraction. It is conceivable that in the intact organ the AII concentration may be much larger in the vicinity of the receptors (biophase) than in the aqueous phase in which the organ is suspended. But the possibility of alterations of the receptor complex attending the preparation of the membranes must be kept in mind.

In view of the above considerations, we believe that, concurrent with the work on receptor isolation, the study of structure-activity relationships will continue to be important for the understanding of the mechanism of action of the hormone. In the case of AII, the discoveries of peptides that act as both reversible (Türker, Yamamoto, Khairallah, and Bumpus, 1971) and irreversible (Paiva, Paiva, Freer, and Stewart, 1972) inhibitors, as well as new data about the phenomenon of tachyphylaxis (Freer and Stewart, 1972; Paiva, Juliano, Nouailhetas, and Paiva, 1974*b*), may contribute to a better understanding of the AII receptors in smooth muscle cells.

TACHYPHYLAXIS OF SMOOTH MUSCLE TO AII

Smooth muscles can be specifically desensitized to AII by repeated treatment with that peptide. This tachyphylactic behavior is thought to be the result of slow dissociation of AII from the receptor, decreasing the number of receptor units available for interaction with new AII molecules (Khairallah, Page, Bumpus, and Türker, 1966). Other mechanisms, such as exhaustion of a transmitter or agonist sequestration by activated storage sites, do not seem to play a significant role in smooth muscle tachyphylaxis to AII (Stewart, 1974).

We have studied tachyphylaxis to AII in two smooth muscle preparations: the isolated guinea pig ileum and the isolated rat uterus. When these organs are suspended in the media normally used for the isolated preparations, tachyphylaxis is observed in the ileum but not in the uterus. However, when the calcium concentration in the medium is lowered from its normal value of 1 mM to about 0.2 mM, the rat uterus can easily be made tachyphylactic to AII (Freer and Stewart, 1972). In the guinea pig ileum, tachyphylaxis is also enhanced by lowering the calcium concentration, as shown in Fig. 2. In this figure the line through the filled circles represents a log dose-

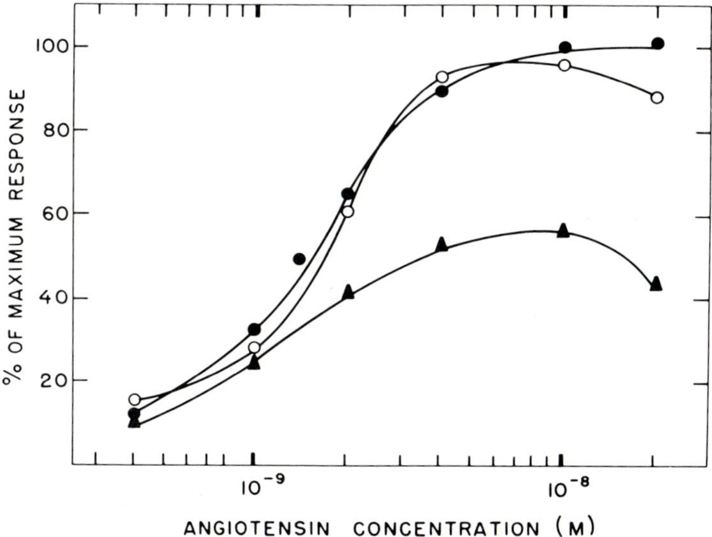

FIG. 2. Log dose-response curves for the action of AII upon the guinea pig ileum in the presence of different Ca^{++} concentrations: ●————●, 1.36 mM; ○————○, 0.68 mM; ▲————▲, 0.34 mM.

response curve obtained at normal calcium concentration (1.36 mM). Increasing doses of AII were administered at 15-min intervals and removed by washing 90 sec after being added (when each contraction reached its maximum). When a similar dose-response experiment was repeated at 0.68 mM Ca^{++} (Fig. 2), some tachyphylaxis was seen with the higher AII doses, as indicated by the downturn of the curve at 2×10^{-8} M AII. When the calcium concentration was further lowered to 0.34 mM (Fig. 2), pronounced tachyphylaxis occurred even with low doses of AII.

Similar dependency of tachyphylaxis on the calcium concentration of the medium has been described for the action of nicotine on rabbit aortic strips (Shibata, Hattori, and Sanders, 1971) and for that of some muscarinic agonists on the isolated guinea pig ileum (Chang and Triggle, 1973). A model for the acetylcholine receptor in guinea pig ileum has been proposed by Chang and Triggle (1973), in which the receptor is normally in a Ca^{++}-associated state (Fig. 3). Upon the reversible interaction of the agonist with

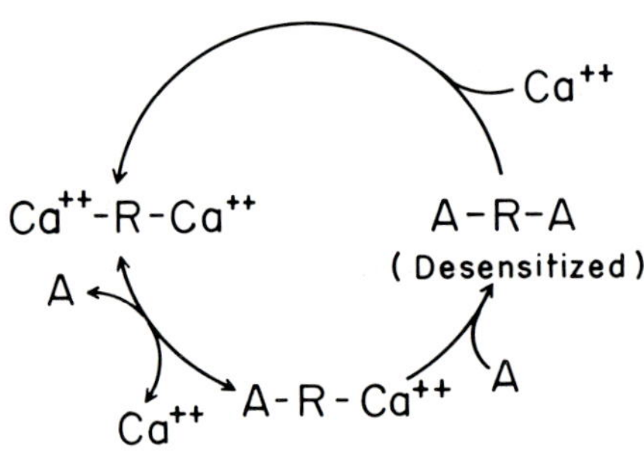

FIG. 3. Our interpretation of the model of Chang and Triggle (1973) for the acetylcholine receptor in guinea pig ileal muscle. A, agonist; R, receptor.

the receptor, the Ca^{++} is dissociated from the latter, which goes into a "permeable" activated state. High agonist concentration would induce a further conformational transition, converting the receptor to a nonpermeable desensitized state, from which a Ca^{++}-dependent recovery to the normal responsive state can take place.

For the case of AII, the tachyphylaxis produced in smooth muscles is fully reversible, since the response can be restored to normal if the organ is left long enough in agonist-free medium. According to Khairallah et al. (1966), this would be due to a very slow dissociation of AII from its receptor. However, there is evidence that the binding site involved in tachyphylaxis is different from the receptor site responsible for the biological response. Some of this evidence comes from the pH dependence of tachyphylaxis.

Freer and Stewart (1972) showed that the tachyphylactic phenomenon in the rat uterus is greatly enhanced by lowering the pH below 7; they interpreted this as an indication that protonation of the histidine side chain in position 6 (His^6) is important for tachyphylaxis. Paiva et al. (1974*b*) confirmed these findings and found a similar effect in the case of the guinea pig ileum. However, the study of a series of AII analogue and homologue peptides, associated with information from electrometric titrations, yielded a good correlation between tachyphylaxis and protonation of the amino group. This is shown in Fig. 4, which contains previously published data (Paiva et al., 1974*b*) with two corrections. Upon further study, the tachyphylaxis of the uterus to Arg^1-AII fell in line with the data for the other peptides, in contrast to the earlier finding of a low degree of tachyphylaxis to that

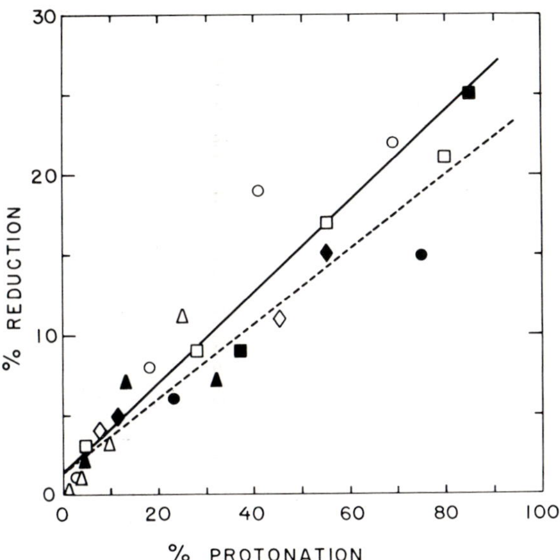

FIG. 4. The effect of amino group protonation on the tachyphylaxis of isolated smooth muscle to AII and four related peptides at several pH values. Tachyphylaxis was measured by the reduction of response to a second maximal dose of the peptide. *Open symbols:* guinea pig ileum; *solid symbols:* rat uterus. *Circles:* AII; *triangles:* Asn^1-AII; *squares:* Gly^1-AII; *lozenges:* Arg^1-AII. The least square lines through the data for the ileum (*solid line:* $y = 1.425 + 0.283x$; $r = 0.95$) and for the uterus (*dashed line:* $y = 1.431 + 0.233x$; $r = 0.95$) are shown. The degree of protonation of peptides at each pH was calculated for 30°C (uterus) and for 37°C (ileum), using the pK and enthalpy values reported by Juliano and Paiva (1974).

peptide. Another correction introduced in Fig. 4 is the calculation of the degrees of protonation at the actual temperatures of the *in vitro* assays (instead of at 25°C, as previously used), made possible by the knowledge of the temperature dependence of the pK values (Juliano and Paiva, 1974).

Figure 4 shows a good correlation between amino group protonation and tachyphylaxis (correlation coefficient, $r = 0.95$ for both the ileum and the uterus data). The correlation of tachyphylaxis with histidine imidazole protonation was poor (Fig. 5), with $r = 0.78$ for the ileum and $r = 0.61$ for

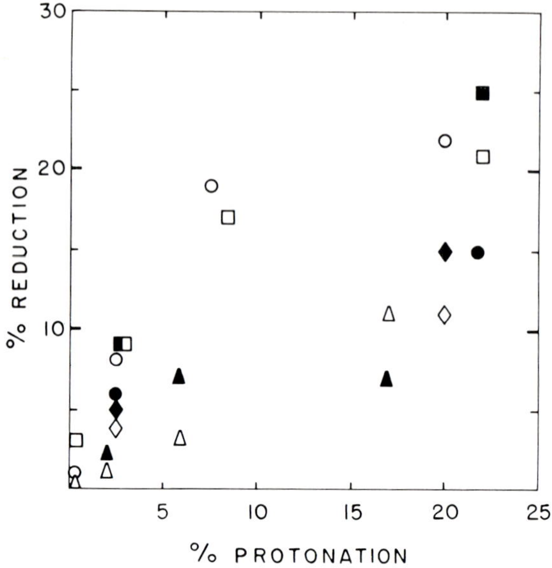

FIG. 5. Tachyphylaxis of the isolated guinea pig ileum (*open symbols*) and rat uterus (*solid symbols*) to AII and related peptides, as a function of the protonation of the imidazole group of the His[6] residue. The symbols are the same as in Fig. 4. The correlation coefficient (*r*) was 0.78 for the ileum and 0.61 for the uterus data.

the uterus. Our conclusion from these observations is that the protonated N-terminal amino group of AII is essential for the production of tachyphylaxis. This conclusion is also supported by the fact that no tachyphylaxis was observed in angiotensin analogues lacking the amino group (desamino-AII) or with a blocked amino group (acetyl-AII). Another conclusion from the data of Fig. 4 is that there is a great similarity between the pH dependence of tachyphylaxis in the ileum and that in the uterus: the two regression lines are not significantly different.

It is interesting to compare the foregoing conclusions regarding the requisites for tachyphylaxis production with the known relationships between structure and agonistic activity of AII. It has long been known that the N-terminal amino group is relatively unimportant for the biological activity of AII. (Paiva and Paiva, 1960). Also, structural changes in the N-terminal end of AII have differing consequences for the agonistic activities in the guinea pig ileum and the rat uterus. An example of this is given in Table 1, where it is shown that on removal of the amino group from AII the activity toward the ileum is decreased 10-fold while that toward the uterus is reduced only about twofold. Table 1 also shows that changes in the Ca^{++} concentration (in the range that was seen to greatly affect tachyphylaxis) do not significantly influence the agonistic activity.

From the above considerations, we conclude that the tachyphylactic

TABLE 1. *Activity of AII and desamino-AII on isolated rat uterus and guinea pig ileum*

Peptide	Guinea pig ileum	$D_{50}{}^{a}$ Rat uterus 1 mM Ca^{++}	0.2 mM Ca^{++}
AII	1.6×10^{-9}	1.8×10^{-9}	1.9×10^{-9}
Desamino-AII	1.5×10^{-8}	3.9×10^{-9}	3.4×10^{-9}

[a] Agonist concentration that produces half-maximal contraction.

phenomenon results from the interaction of the protonated amino group with a site on the cell membrane that is not a part of the receptor responsible for the agonistic activity of AII. This tachyphylactic site might be an allosteric site with a lower affinity for AII than the receptor but with a much smaller dissociation rate constant. Binding of AII to the site would change the AII-receptor complex into a desensitized state. It is also possible that tachyphylactic site occupation would lead to decreased affinity of the receptor for AII. However, tachyphylaxis is mainly a reduction of the organ's maximum response to AII, although some apparent loss of affinity seems also to be present (Paiva, Mendes, and Paiva, *unpublished results*).

The hypothesis that tachyphylaxis results from AII binding to an allosteric site in a reversible manner, together with its enhancement by Ca^{++}, may be represented as the following modification of Chang and Triggle's (1973) model:

$$Ca^{++}\!\!-\!\!Rec\!\!-\!\!Ca^{++} \underset{Ca^{++}}{\overset{A}{\rightleftharpoons}} Ca^{++}\!\!-\!\!Rec\!\!-\!\!A \underset{Ca^{++}}{\overset{A}{\rightleftharpoons}} A\!\!-\!\!Rec\!\!-\!\!A$$

Another possibility is that the tachyphylactic site is close to the specific binding sites of the receptor, as proposed by Stewart (1974). This possibility is favored by the strict structural specificity observed in the tachyphylactic phenomenon. It is known, for instance, that des-Asp[1]-AII, which has considerable agonistic activity, does not induce tachyphylaxis, even when its amino group is protonated (Paiva et al., 1974*b*). It is also known that replacement of His[6] in AII by a pyrazolylalanine or a thienylalanine residue also yielded analogues incapable of producing tachyphylaxis (Freer and Stewart, 1972). This indicates that the interaction of AII with the tachyphylactic site depends also on binding to the specific receptor sites. For this reason we propose the model depicted in Fig. 6, in which the tachyphylactic site is near the AII receptor. When the amino group binds to its receptor site, the receptor is activated (Fig. 6A); but it may also bind to the tachyphylactic site, in competition with calcium ions (Fig. 6B). When this hap-

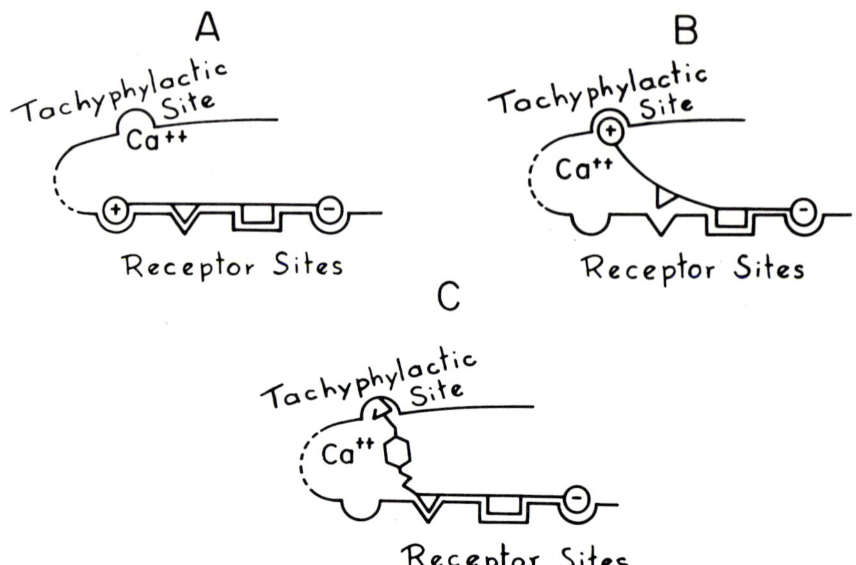

FIG. 6. Model proposed for the relation between the AII receptor and the tachyphylactic site. A: The AII molecule binds to the receptor through the C-terminal end (negative sign), the N-terminal end (positive sign), and other unspecified groups. B: When the amino group is bound to the tachyphylactic site, the receptor is inactivated. C: Covalent binding of Chl[1]-AII to the tachyphylactic site.

pens, the fit of one or more essential groups of AII to the receptor may be altered, leaving the hormone-receptor complex in an inactive state. This receptor model is a modification of the one proposed by Freer and Stewart (1972), in which the tachyphylactic site interacts with the imidazole group of the histidine side chain in position 6 of AII.

AFFINITY LABELING OF AII RECEPTORS

In an attempt to label AII receptors irreversibly, Paiva et al. (1972) prepared p-|N,N-bis(2-chloroethyl)amino|phenylbutyryl (chlorambucil) derivatives of AII, in the hope of alkylating some nucleophilic group in or near the receptor through the nitrogen mustard chlorambucil moiety. Chlorambucil[1]-angiotensin II (Chl[1]-AII), in which the alkylating moiety replaces the N-terminal Asp residue of AII, was shown to be an antagonist of AII action on guinea pig ileum, and the antagonism was of the noncompetitive irreversible type. In the rat uterus, the antagonism was not observed when the organ was suspended in a medium containing 1 mM Ca++. However, an antagonism comparable to that seen in the guinea pig ileum was observed when the treatment with Chl[1]-AII was done at 0.18 mM Ca++, and it was more intense when the treatment was done in a calcium-free medium (Paiva, Miyamoto, and Paiva, 1974*c*). Low Ca^{++} concentrations also enhanced the inhibition observed in the guinea pig ileum.

The Ca^{++} dependence of the Chl^1-AII antagonism in the rat uterus and guinea pig ileum is very similar to the Ca^{++} dependence of tachyphylaxis, suggesting that the same site might be involved in both phenomena. Accordingly, we suggest that Chl^1-AII exerts its irreversible inhibition of the action of AII on the rat uterus and the guinea pig ileum through alkylation of the same site involved in tachyphylaxis (Fig. 6C). If this hypothesis holds, it should be kept in mind when Chl^1-AII is used as an affinity label for the purification of AII receptors, since during the purification steps, the specific receptor sites might separate from the labeled tachyphylactic site.

ACKNOWLEDGMENTS

The authors' research was supported by grants from the São Paulo State Research Foundation (FAPESP), the Brazilian National Research Council (CNPq), Financiadora de Estudos e Projetos, S.A. (FINEP), and "Central de Medicamentos" (CEME).

REFERENCES

Baudouin, M., Meyer, P., Fermandjian, S., and Morgat, J. L. (1972): Calcium release induced by interaction of angiotensin with its receptors in smooth muscle cell microsomes. *Nature* 235:336–338.

Chang, K. J., and Triggle, D. J. (1973): Quantitative aspects of drug-receptor interactions. II. The role of Ca$^{++}_{mem}$ in densensitization and spasmolytic activity. *J. Theor. Biol.* 40:155–172.

Devynck, M. A., Pernollet, M. G., Meyer, P., Fermandjian, S., and Fromageot, P. (1973): Angiotensin receptors in smooth muscle cell membranes. *Nature, New Biol.* 245:55–58.

Devynck, M. A., Pernollet, M. G., Meyer, P., Fermandjian, S., Fromageot, P., and Bumpus, F. M. (1974): Solubilisation of angiotensin II receptors in rabbit aortae membranes. *Nature* 249:67–69.

Freer, R. J., and Stewart, J. M. (1972): Some characteristics of uterine angiotensin receptors. In: *Structure-Activity Relationships of Proteins and Polypeptide Hormones,* edited by M. Margoulies and F. C. Greenwood, pp. 490–495. Excerpta Medica, Amsterdam.

Juliano, L., and Paiva, A. C. M. (1974): Conformation of angiotensin II in aqueous solution. Titration of several peptide analogs and homologs. *Biochemistry* 13:2445–2450.

Khairallah, P. A., Page, I. H., Bumpus, F. M., and Türker, R. K. (1966): Angiotensin tachyphylaxis and its reversal. *Circ. Res.* 19:247–254.

Oliveira, M., and Holzhacker, S. (1974): Isolation and characterization of smooth muscle cell membranes. *Biochim. Biophys. Acta* 332:221–232.

Paiva, A. C. M., and Paiva, T. B. (1960): The importance of the N-terminal end of angiotensin II for its pressor and oxytocic activities. *Biochem. Pharmacol.* 5:187–191.

Paiva, T. B., Goissis, G., Juliano, L., Miyamoto, M. E., and Paiva, A. C. M. (1974*a*): Angiotensin-like and antagonistic activities of N-terminal modified |8-Leucine|angiotensin II peptides. *J. Med. Chem.* 17:238–241.

Paiva, T. B., Juliano, L., Nouailhetas, V. L. A., and Paiva, A. C. M. (1974*b*): The effect of pH on tachyphylaxis to angiotensin peptides in the isolated guinea pig ileum and rat uterus. *Eur. J. Pharmacol.* 25:191–196.

Paiva, T. B., Miyamoto, M. E., and Paiva, A. C. M. (1974*c*): Irreversible inhibition of angiotensin II activity on the rat uterus by an alkylating angiotensin derivative. *Pharmacology,* 12:186–191.

Paiva, T. B., Paiva, A. C. M., Freer, R. J., and Stewart, J. M. (1972): Alkylating analogs of peptide hormones. 2. Synthesis and properties of p-|N,N-bis(2-chloroethyl)amino|phenyl-butyryl derivatives of angiotensin II. *J. Med. Chem.* 15:6–8.

Rasmussen, H. (1972): Cell receptor sites for peptide hormones. In: *Structure-Activity Relationships of Protein and Polypeptide Hormones,* edited by M. Margoulies and F. C. Greenwood, pp. 194–198. Excerpta Medica, Amsterdam.

Schreier-Muccillo, S., Niculitcheff, G. X., Oliveira, M. M., Shimuta, S., and Paiva, A. C. M. (1974): Conformational changes at membranes of target cells induced by the peptide hormone angiotensin. A spin label study. *FEBS Lett.,* 47:193–196.

Shibata, S., Hattori, K., and Sanders, B. (1971): The relationship between external calcium concentration and the recovery rate of aortic strips from nicotine tachyphylaxis. *Eur. J. Pharmacol.* 16:109–112.

Somlyo, A. P., and Somlyo, A. V. (1968): Vascular smooth muscle. I, Normal structure, pathology, biochemistry and biophysics. *Pharmacol. Rev.* 20:197–272.

Stewart, J. M. (1974): Tachyphylaxis to angiotensin. In: *Angiotensin,* edited by I. H. Page and F. M. Bumpus, pp. 170–184. Springer-Verlag, Berlin.

Türker, R. K., Yamamoto, M., Khairallah, P. A., and Bumpus, F. M. (1971): Competitive antagonism of 8-Ala-angiotensin II to angiotensin I and II on isolated rabbit aorta and rat ascending colon. *Eur. J. Pharmacol.* 15:285–291.

Concepts of Membranes in Regulation and Excitation,
edited by M. Rocha e Silva and G. Suarez-Kurtz.
Raven Press, New York © 1975.

Permeability Characteristics of Lipid Bilayers Revealed by Spin Probes

Shirley Schreier-Muccillo, D. Marsh,* and I. C. P. Smith[†]

*Department of Biochemistry, Institute of Chemistry, University of São Paulo, São Paulo, S.P. Brazil, and *Department of Biochemistry, University of Oxford, Oxford, England, and †Division of Biological Sciences, National Research Council, Ottawa, Ontario, Canada*

INTRODUCTION

Membrane Permeability

Biological membranes participate in the qualitative and quantitative control of the flow of materials inside and outside cells. This delicate balance can be achieved only through specific membrane organization. An understanding of the arrangement of membrane components and how their molecular structure contributes to that specificity is highly desirable.

Lipid bilayers (Henn and Thompson, 1969) have been widely used to mimic the lipid portion of biological membranes, since it has been shown (Blaurock and Wilkins, 1969; Engelman, 1971; Esfahafani, Limbrick, Knutton, Oka, and Wakil, 1971) that a large lipid fraction of certain membranes occurs in bilayers.

The phenomenon of permeation of solutes—both ionic and nonionic—and of water has been dealt with theoretically by different approaches. Models based on irreversible thermodynamics (Kedem and Katchalsky, 1958) and explanations at a molecular level (Lieb and Stein, 1971) have been proposed.

A large body of experimental data concerning permeability properties of biological (Diamond and Wright, 1969) and artificial membranes (Bangham, 1968) has been accumulated.

Numerous methods—such as diffusion of radioactive permeants (Papahadjopoulos and Bangham, 1966); electrical (Tien and Diana, 1968), optical (Bangham, de Gier, and Greville, 1967), and surface pressure measurements (Phillips, 1972); nuclear magnetic resonance (nmr) spectra (Chapman, 1972); and spin-probe, or electron spin resonance (esr), spectra (Schreier-Muccillo and Smith, 1974)—have been used for a variety of lipid systems—organized either as monolayers (Phillips, 1972), planar single layers (Mueller, Rudin, Tien, and Wescott, 1963) or multibilayers (Levine, Bailey, and Wilkins, 1968), or as single (Huang, 1969) or multibilayer vesicles (Bangham, 1968) — in order to try to relate behavior to permeability properties.

It has been found that permeability characteristics depend on the composition of the head group (polar region) of the phospholipid constituents of a given system and on the degree of fluidity of the hydrophobic core (van Deenen, 1972). The latter has been shown to depend on the length and degree of unsaturation of the acyl chains of phospholipids and on the presence of compounds, such as sterols, able to interact with them (van Deenen, 1972).

Some studies (de Gier, Mandersloot, Hupkes, McElhaney, and van Beek, 1971; Kroes and Ostwald, 1971; Graziani and Livne, 1972; McElhaney, de Gier, and van der Neut-Kok, 1973) report on the activation energies of the permeation process, and it has been suggested that in some cases (de Gier et al., 1971; Kroes and Ostwald, 1971; McElhaney et al., 1973) the removal of water of hydration is the rate-determining step for the partitioning of a solute into the membrane.

Use of Spin Probes in the Study of Membrane Structure

The spin-label method (Hamilton and McConnell, 1968) has been used extensively in the study of biological and model membranes (Schreier-Muccillo and Smith, 1974). The spectral parameters — hyperfine splittings, g-value, line widths, and line shapes — are those commonly employed to gain information concerning the degree of motion and orientation of spin probes in membranes. The spectral effects of intermolecular interaction (spin-spin interaction) have been used to calculate the coefficient of lateral diffusion (Devaux and McConnell, 1972; Sackmann and Träuble, 1972a,b; Scandella, Devaux, and McConnell, 1972; Träuble and Sackmann, 1972; Sackmann, Träuble, Galla, and Overath, 1973; Stier and Sackmann, 1973) and intermolecular distances (Marsh and Smith, 1972, 1973) in membranes. Spin-spin interaction has also given information about the degree of packing in lipid systems (Schreier-Muccillo, Butler, and Smith, 1973).

The reduction of the nitroxide group of spin probes by ascorbic acid is a known fact (McConnell and McFarland, 1970). It has been used to demonstrate migration (flip-flop) of phospholipids from the inner to the outer monolayer of vesicles, and vice versa (Kornberg and McConnell, 1971). In that work the probe containing the nitroxide moiety in the polar head group was

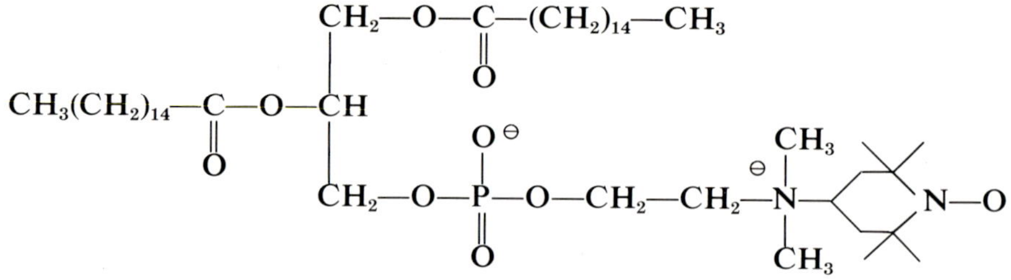

$$CH_2\!-\!O\!-\!\underset{\underset{\textstyle O}{\|}}{C}\!-\!(CH_2)_{14}\!-\!CH_3$$

$$CH_3(CH_2)_{14}\!-\!\underset{\underset{\textstyle O}{\|}}{C}\!-\!O\!-\!CH$$

$$CH_2\!-\!O\!-\!\underset{\underset{\textstyle O}{\|}}{\overset{\overset{\textstyle O^{\ominus}}{|}}{P}}\!-\!O\!-\!CH_2\!-\!CH_2\!-\!\overset{\ominus}{N}\!\!\begin{smallmatrix}CH_3\\|\\ \\|\\CH_3\end{smallmatrix}\!\!-\!\!\big\langle\;\big\rangle\!-\!N\!-\!O$$

employed, and its reduction by ascorbate in the aqueous phase at O°C (when the permeability of the membrane to the ion is very low) was studied.

Here we report on the reduction by ascorbate of the nitroxide groups of spin probes located *within* the membrane (Fig. 1), except for probe I, as will be seen later. The kinetics of this process allowed us to confirm ideas about the relative location of the rings containing the paramagnetic group in the bilayers, based on estimated values of the isotropic hyperfine splittings (Seelig, 1970). The dependence of the rate of reaction and activation energy on the head-group composition and degree of packing of the hydrophobic portion of the lipid bilayers led us to propose a model for the permeation process based on a partition-diffusion mechanism (Lieb and Stein, 1971).

The kinetics of reduction of probes by ascorbate was later used by Butler (1974) to examine the interaction between local anesthetics and lipid membranes.

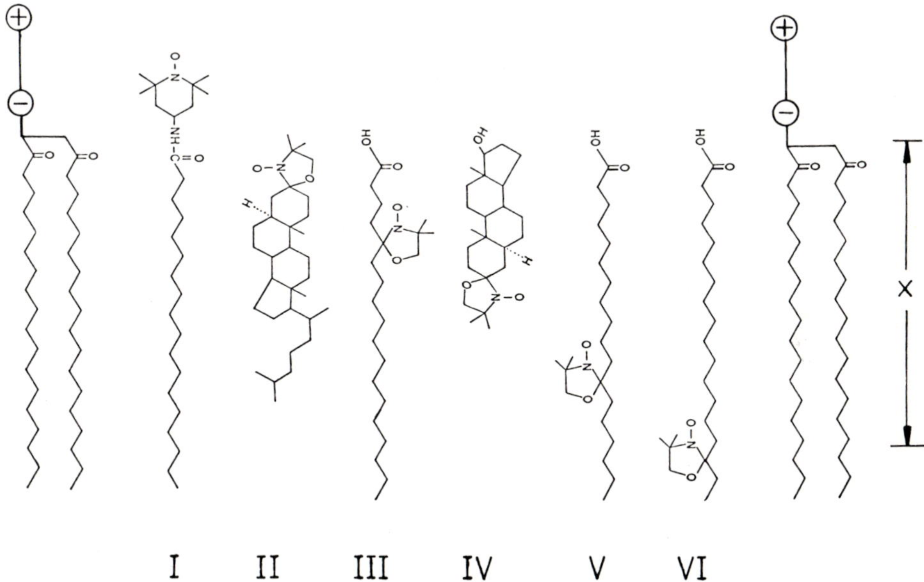

I   II   III   IV   V   VI

FIG. 1. Spin probes used in the present study. The relative position of the nitroxide-containing rings in the bilayer is indicated (see text).

PREPARATION OF SAMPLES

Planar lipid multibilayers were prepared on the inner surfaces of flat aqueous standard esr cells by blowing wet nitrogen through an organic solvent solution containing the desired components. The concentration of the spin probe was 0.7% (in moles) of total lipid.

The samples were dried under vacuum. The cells were mounted in the cavity of a Varian E-9 spectrometer, equipped for variable temperature work. Butler (1972, *unpublished*) designed the portion of the equipment that fitted the cell in the cavity. The cells were mounted with the plane of the bilayers perpendicular to the applied magnetic field.

Samples were allowed to equilibrate at a given temperature, and solutions of ascorbate (pH 6.2 to 6.8), previously equilibrated at the same temperature, were added.

REDUCTION OF SPIN PROBES BY ASCORBATE

The Probes and Systems Used

Figure 1 shows the probes used in the present study. They are derivatives of stearic acid and steroids, analogues of intrinsic membrane components.

We have used the systems PC (egg phosphatidylcholine) and PC–35 mole % Chol (cholesterol) to examine the effect of fluidity of the hydrophobic core of the bilayer on the rate of reduction (see below). Other systems were PC (35 mole %)–30 mole % PS (beef brain phosphatidyl serine)–35 mole % Chol and 50 mole % PC–15 mole % HDTMA (hexadecyl trimethyl ammonium chloride)–35 mole % Chol which were compared to PC–35 mole % Chol to analyze the effect of head-group composition. It was assumed that the same cholesterol content would provide a similar degree of packing in the lipophilic region of the membrane.

THE KINETICS OF REACTION

Half-times

In contrast to what is the very fast reduction of paramagnetic species observed for the reaction of ascorbate with the nitroxide group in aqueous solution (Kornberg and McConnell, 1971)—50 μl of a 3.3 10^{-3} M solution of the label had the amplitude of its spectrum decreased by 99.9% in 2 min at 0°C upon the addition of 5 μl of 0.35 M sodium ascorbate, pH 7.5—the rate of disappearance of the esr signal of the lipid probes intercalated in the bilayers is slow and was seen to depend on the molecular structure of the probe and the lipid composition of the bilayer.

The decay of the heights of the esr lines was used to monitor the process, since no change occurred in the line shape during the time course of the experiments. Figure 2 illustrates one such experiment. Spectra were taken at definite intervals after shifting the field by 1 Gauss each time. The exponential decay is clear. Figure 3 is a plot of the logarithm of the heights of the lines versus time. It shows that the process follows first-order kinetics with respect to spin probe. The same was observed for ascorbate.

Table 1 lists the half-times for reaction at 19.2°C for all systems studied.

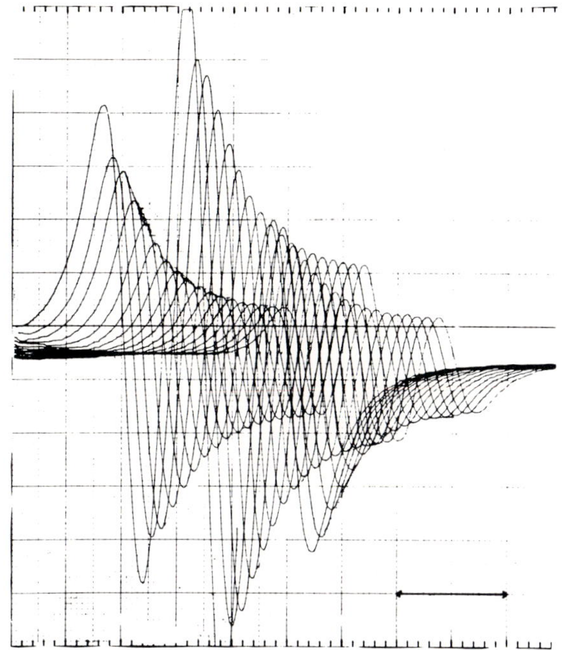

FIG. 2. The esr spectra of probe II in oriented multibilayers of PC (50%), HDTMA (15%), and Chol (35%) as a function of time after hydration with 10^{-3} M sodium ascorbate at pH 6.5, 6.0°C. The first spectrum (*left*) was taken immediately after hydration and the others at regular intervals. For each measurement the starting point of the magnetic field sweep was stepped by 1 Gauss. Spectra were taken with the applied magnetic field perpendicular to the plane of the film.

The half-time at this temperature could not be determined for probe I, since the reaction was very fast. At 3.5°C, half-time was 3 min in PC–35 mole % Chol. This should be compared to the value of 13 min at 4.0°C for probe II. These results and those in Table 1 indicate a decreased accessibility of the nitroxide to ascorbate when the former is located further away from the polar region of the bilayer—according to expectations based on molecular structure and on estimated values of isotropic hyperfine splittings (Seelig, 1970).

Table 1 also shows that for a given probe the process is slower if the hydrophobic core is more tightly packed (the half-time for PC–Chol is larger than the half-time for PC alone); the half-time is also larger for the membranes containing PS (negatively charged) than for those containing PC–Chol (no net charge), which, in turn, is larger than that for the system containing HDTMA (positively charged). This, again, is in agreement with expectations, taking into account that ascorbate bears one negative charge at the pH of the experiments ($pK_{a_1} = 4.2$).

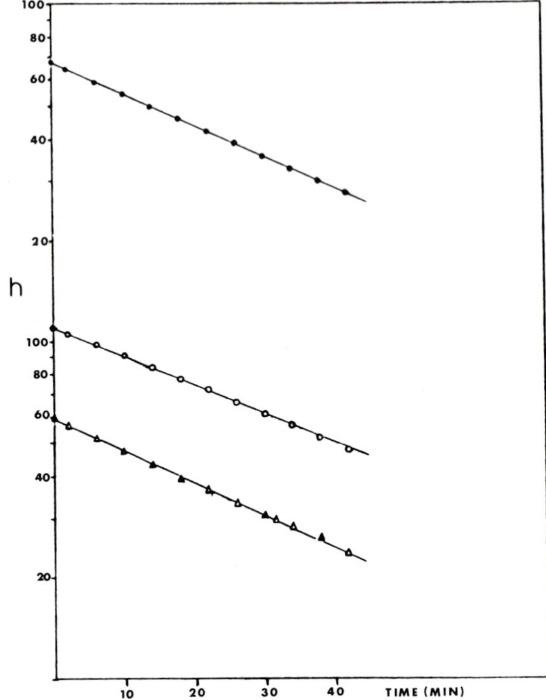

FIG. 3. Plots of the logarithm of line intensities for the esr spectra of probe II in hydrated multibilayers of PS (30%), PC (35%) and Chol (35%) hydrated with 10^{-2} M sodium ascorbate, pH 6.6, 8.3°C as a function of time after hydration. The three curves are for the low field (h_1, $\bigcirc$), central (h_0, $\bullet$) and high field (h_{-1}, $\triangle$) of the spectra taken with the magnetic field perpendicular to the plane of the bilayer film.

TABLE 1. *Half-times (min) at 19.2°C for the reaction between ascorbate (0.01 M, pH 6.2–6.8) and spin probes intercalated in lipid bilayers*

Spin probe	PC	PC–35% Chol	35% PC–30% PS–35% Chol	50% PC–15% HDTMA–35% Chol
II	—	3.2	12	5.9[a]
III	9.5	11.5	51	—
IV	16	29	85	16
V	30	60	136	—
VI	32	—	166	34

[a] [Ascorbate] = 0.001 M.

Activation Energies

The activation energies were calculated from variable temperature studies and are given in Table 2. It is seen that the activation energy did not vary significantly with the nature of the probe or the head-group composition. A

TABLE 2. *Activation energies* (E_a) *for the reaction between ascorbate (0.01 M, pH 6.2–6.8)* and spin probes intercalated in lipid bilayers

Spin probe	PC	PC–35% Chol	35% PC–30% PS–35% Chol	50% PC–15% HDTMA–35% Chol
			E_a (kcal/mol)	
II	—	14	15	15
III	6	13	—	—
IV	6	16	14	16
V	6	15	—	—

considerable change was observed, however, when the fluidity of the hydrocarbon core was changed.

DISCUSSION

Location of Probes

The results obtained in Table 1 allow us to draw some conclusions about the relative position of the nitroxide-containing rings of probes I to VI in the bilayers, as depicted in Fig. 1.

It had been previously shown (Kornberg and McConnell, 1971) that a water-soluble spin label and the head-group spin-labeled phospholipid underwent fast reduction by ascorbate. Our results with probe I, whose paramagnetic moiety is probably exposed to the water-membrane interface, are consistent with the previous observations.

It should be noted that although probe II is reduced much faster than probes III to VI, its half-time of reduction is considerably larger than that of probe I under similar conditions. This indicates that the nitroxide moiety of probe II, although considerably exposed to the aqueous interface, is more buried than that of probe I and therefore not located in the head-group region of the membrane. This should be taken into account when spectral features of different probes are being analyzed. For example, in a study of the effect of cholesterol on egg PC and dipalmitoyl PC, Schreier-Muccillo, Marsh, Dugas, Schneider, and Smith (1973) found quantitatively different results between the stearic acid probes III and V and the steroid probes II and IV. The results were ascribed to differences in molecular shape and rigidity of the probes, and not to the fact that they were probing different regions of the bilayer. This explanation finds further support in the present results.

Permeability Characteristics of Lipid Bilayers. Mechanism of Permeation

It was mentioned earlier that the rate of reaction of ascorbate with spin labels in aqueous solutions is very fast. Assuming that the polarity of the environment does not significantly affect the rate of a radical reaction, we ascribe the results on lipid bilayers to the process of permeation.

A mechanism has been proposed (Lieb and Stein, 1971) by which the permeation of a solute through a membrane would depend on its partition coefficient into the membrane (K_p), on the diffusion coefficient (D) within the membrane, and on the membrane thickness (l). The permeability coefficient is given by

$$P = \frac{K_p D}{l} (l) \tag{1}$$

Our experimental results can be interpreted according to Eq. 1. In this sense, the charges on the membrane surface (positive when HDTMA is present, negative when PS is present) will attract or repel the ascorbate ion, increasing or decreasing its effective concentration in the diffuse double layer, as compared to the system containing only PC with no net charge — all three in the presence of 35 mole % Chol. This will affect the partitioning of ascorbate into the membrane. Table 1 shows that the half-times of reduction for a given probe increase in the expected manner in the above systems.

An examination of the activation-energy values (Table 2) reveals that these are only slightly affected by the variation in head-group composition. This suggests that the partitioning step does not contribute significantly to the activation energy of the overall process; that is, it is not the rate-determining step for permeation of ascorbate. Indeed, the contribution to the activation energy of the surface potential due to the charged lipids has been calculated (Schreier-Muccillo, Marsh, and Smith, *submitted*). It was seen that the charge effect would affect the activation energy by an amount comparable to the experimental error.

De Gier et al. (1971), Kroes and Ostwald (1971), Graziani and Livne (1972), and McElhaney et al. (1973) have not observed changes in activation energy for permeation of a series of nonionic solutes and water through a variety of membranes when the cholesterol content was changed. Therefore, in the case of the nonionic solutes, it was proposed that the rate-determining step would be the partitioning of the solute into the membrane. The removal of water of hydration for membrane penetration by such compounds was suggested as the main factor determining the energetics of partition. In contrast to those results, a variation of the cholesterol content in the systems studied by us caused changes in both the half-times and activation energies (Tables 1 and 2) for the reduction of the spin probes by ascorbate.

In the foregoing discussion about the effect of head groups it was assumed that the fluidity of the hydrocarbon portion of the membranes containing 35 mole % Chol is similar. A much lower degree of packing is achieved by PC alone (Lapper, Paterson, and Smith, 1972; Schreier-Muccillo et al., 1973). When the two systems, PC and PC–35 % Chol, are compared, it is seen (Tables 1 and 2) that both the half-times and activation energies vary for a given label. These results suggest that the diffusion of ascorbate in the membrane plays an important role in the overall rate process, contributing significantly to the activation energy. Thus diffusion in the hydrophobic core of the bilayer constitutes a rate-determining step in the process of permeation of ascorbate.

That agrees with the role postulated by Lieb and Stein (1971) for diffusion in the mechanism of permeation. A more quantitative analysis of the results has been carried out (Schreier-Muccillo et al., *submitted*), and the surface potential values for bilayers containing PS and HDTMA have been calculated. Permeability coefficients have been estimated.

Application of the Present Work

Most experiments dealing with permeability of membranes are based on the measurement of macroscopic properties, usually performed in the aqueous region surrounding the membrane (Papahadjopoulos and Ohki, 1970).

The use of spin probes with the paramagnetic moiety located in different regions of lipid bilayers provides a convenient means of analyzing the permeation process *within* the bilayer at varying distances from the aqueous interface.

The principle involved in the present experiments could easily be extended to other techniques, thus introducing a wide range of possibilities for the study of the process of permeation at the molecular level.

ACKNOWLEDGMENTS

S.S.-M. is grateful to the São Paulo State Research Foundation (FAPESP) for partial support of this work. We are grateful to J. Aboulafia, N. Miasiro and S. Shimuta for comments on the manuscript.

REFERENCES

Bangham, A. D. (1968): Membrane models with phospholipids. *Progr. Biophys. Mol. Biol.* 18:29–95.

Bangham, A. D., de Gier, J., and Greville, G. D. (1967): Osmotic properties and water permeability of phospholipid liquid crystals. *Chem. Phys. Lipids* 1:225–246.

Blaurock, A. E., and Wilkins, M. H. F. (1969): Structure of frog photoreceptor membranes. *Nature* 223:906–909.

Butler, K. W. (1974): The effect of anesthetics on membrane lipids—A spin probe study. *J. Pharm. Sci. In press.*

Chapman, D. (1972): Nuclear magnetic resonance spectroscopic studies of biological membranes. *Ann. N.Y. Acad. Sci.* 195:179–201.

de Gier, J., Mandersloot, J. G., Hupkes, J. V., McElhaney, R. N., and van Beek, W. P. (1971): On the mechanism of non-electrolyte permeation through lipid bilayers and through membranes. *Biochim. Biophys. Acta,* 233:610–618.

Devaux, P., and McConnell, H. M. (1972): Lateral diffusion in spin-labeled phosphatidyl choline multilayers. *J. Am. Chem. Soc.* 94:4475–4481.

Diamond, J. M., and Wright, E. M. (1969): Biological membranes: The physical basis of ion and nonelectrolyte selectivity. *Annu. Rev. Physiol.* 31:581–646.

Engelman, D. M. (1971): Lipid bilayer structure in the membrane of *Mycoplasma laidlawii*. *J. Mol. Biol.* 58:153–165.

Esfahafani, M., Limbrick, A. R., Knutton, S., Oka, T., and Wakil, S. J. (1971): The molecular organization of lipids in the membrane of *Escherichia coli:* Phase transitions. *Proc. Nat. Acad. Sci. U.S.A.* 68:3180–3184.

Graziani, Y., and Livne, A. (1972): Water permeability of bilayer lipid membranes: Sterol-lipid interaction. *J. Membr. Biol.* 7:275–284.

Hamilton, C. L., and McConnell, H. M. (1968): Spin labels. In: *Structural Chemistry and Molecular Biology*, edited by A. Rich and N. Davidson, pp. 115–149. W. H. Freeman and Co., San Francisco.

Henn, F. A., and Thompson, T. E. (1969): Synthetic lipid bilayer membranes. *Annu. Rev. Biochem.* 38:241–262.

Huang, C. (1969): Studies on phosphatidylcholine vesicles. Formation and physical characteristics. *Biochemistry* 8:344–352.

Kedem, O., and Katchalsky, A. (1958). Thermodynamic analysis of the permeability of biological membranes to non-electrolytes. *Biochim. Biophys. Acta,* 27:229–246.

Kornberg, R. D., and McConnell, H. M. (1971): Inside-outside transitions of phospholipids in vesicle membranes. *Biochemistry* 10:1111–1120.

Kroes, J., and Ostwald, R. (1971): Erythrocyte membranes-Effect of increased cholesterol content on permeability. *Biochim. Biophys. Acta* 249:647–650.

Lapper, R. D., Paterson, S. J., and Smith, I. C. P. (1972): A spin label study of the influence of cholesterol on egg lecithin multibilayers. *Can. J. Biochem.* 50:969–981.

Levine, Y. K., Bailey, A. I., and Wilkins, M. H. F. (1968): Multilayers of phospholipid bimolecular leaflets. *Nature* 220:577–578.

Lieb, W. R., and Stein, W. D. (1971): The molecular basis of simple diffusion within biological membranes. In: *Current Topics in Membranes and Transport,* edited by F. Bronner and A. Kleinzeller, pp. 1–39. Academic Press, New York.

Marsh, D., and Smith, I.C.P. (1972): Interacting spin labels as probes of molecular separation within phopholipid bilayers. *Biochem. Biophys. Res. Commun.* 49:916–922.

Marsh, D., and Smith, I. C. P. (1973): An interacting spin label study of the fluidizing and condensing effects of cholesterol on lecithin bilayers. *Biochim. Biophys. Acta* 298:133–144.

McConnell, H. M., and McFarland, B. G. (1970): Physics and chemistry of spin labels. *Q. Rev. Biophys.* 3:91–136.

McElhaney, R. N., de Gier, J., and van der Neut-Kok, E. C. M. (1973): The effect of alterations in fatty acid composition and cholesterol content on the nonelectrolyte permeability of *Acholeplasma laidlawii* B. cells and derived liposomes. *Biochim. Biophys. Acta* 298:500–512.

Mueller, P., Rudin, D. O., Tien, H. T., and Wescott, W. C. (1963): Methods for the formation of single bimolecular lipid membranes in aqueous solution. *J. Phys. Chem.* 67:534–535.

Papahadjopoulos, D., and Bangham, A. D. (1966): Biophysical properties of phospholipids. II. Permeability of phosphatidyl serine liquid crystals to univalent ions. *Biochim. Biophys. Acta* 126:185–188.

Papahadjopoulos, D., and Ohki, S. (1970): Conditions of stability for liquid crystalline phospholipid membranes. In: *Liquid Crystals and Ordered Fluids,* edited by J. F. Johnston and R. S. Porter, pp. 13–32. Plenum Press, New York.

Phillips, M. C. (1972): The physical state of phospholipids and cholesterol in monolayers, bilayers, and membranes. In: *Progress in Surface and Membrane Science,* edited by J. F. Danielli, M. D. Rosenberg, and D. A. Cadenhead, Vol. 5, pp. 139–221. Academic Press, New York.

Sackmann, E., and Träuble, H. (1972*a*): Studies of the crystalline-liquid crystalline phase transition of lipid model membranes. I. Use of spin labels and optical probes as indicators of the phase transition. *J. Am. Chem. Soc.* 94:4482–4491.

Sackmann, E., and Träuble, H., (1972*b*): Studies of the crystalline-liquid crystalline phase transition of lipid model membranes. II. Analysis of electron spin resonance spectra of steroid labels incorporated into lipid membranes. *J. Am. Chem. Soc.* 94:4492–4498.

Sackmann, E., Träuble, H., Galla, H-J., and Overath, P. (1973): Lateral diffusion, protein mobility, and phase transitions in *Escherichia coli* membranes. A spin label study. *Biochemistry* 12:5360–5369.

Scandella, C. J., Devaux, P., and McConnell, H. M. (1972): Rapid lateral diffusion of phospholipids in rabbit sarcoplasmic reticulum. *Proc. Nat. Acad. Sci. U.S.A.* 69:2056–2060.

Schreier-Muccillo, S., Butler, K. W., and Smith, I. C. P. (1973): Structural requirements for the formation of ordered lipid multibilayers. *Arch. Biochem. Biophys.* 159:297–311.

Schreier-Muccillo, S., Marsh, D., Dugas, H., Schneider, S., and Smith, I. C. P. (1973): A spin probe study of the influence of cholesterol on the motion and orientation of phospholipids in oriented multibilayers and vesicles. *Chem. Phys. Lipids* 10:11–17.

Schreier-Muccillo, S., and Smith, I. C. P. (1974): *Spin Labels as Probes of Biological Membranes*. Academic Press, New York.

Seelig, J. (1970): Spin label studies of oriented smectic liquid crystals (a model system for bilayer membranes). *J. Am. Chem. Soc.* 92:3881–3887.

Stier, A., and Sackmann, E. (1973): Spin labels as enzyme substrates-Heterogeneous lipid distribution in liver microsomal membranes. *Biochim. Biophys. Acta* 311:400–408.

Tien, H. T., and Diana, A. L. (1968): Bimolecular lipid membranes: A review and summary of some recent studies. *Chem. Phys. Lipids* 2:55–101.

Träuble, H., and Sackmann, E. (1972): Studies of the crystalline-liquid crystalline phase transition of lipid model membranes. III. Structure of a steroid-lecithin system below and above the lipid-phase transition. *J. Am. Chem. Soc.* 94:4499–4510.

van Deenen, L. L. M. (1972): Phospholipide-Beziehungen zwischen ihrer chemischen Struktur und Biomembranen. *Naturwissenschaften* 59:485–491.

Concepts of Membranes in Regulation and Excitation,
edited by M. Rocha e Silva and G. Suarez-Kurtz.
Raven Press, New York © 1975.

Angiotensin Receptors in Smooth Muscle Cell Membranes. A Spin-label Study

Mecia M. Oliveira, Shirley Schreier-Muccillo,* Suma Shimuta, Gregory Niculitcheff,† and A. C. M. Paiva

*Department of Biophysics and Physiology, Escola Paulista de Medicina, 04023 São Paulo, S.P., Brazil, and * Department of Biochemistry, Institute of Chemistry, University of São Paulo, São Paulo, S.P., Brazil, and †Department of Biochemistry, Faculty of Medicine of Santa Casa de Misericórdia, São Paulo, S.P., Brazil*

INTRODUCTION

Hormone-receptor interaction is the first step in a chain of cellular processes leading to hormonal responses. Receptor sites are believed to be macromolecular components, finite in number, possessing a genetically determined structure, and capable of specific recognition of and interaction with hormones with a high degree of specificity and affinity. Receptors of steroid hormones are intracellular, whereas those of neurotransmitters and polypeptide hormones are located in the cell membranes (Rasmussen, 1971).

Angiotensin II (AII) is a polypeptide hormone with, among many pharmacological actions, the basic ability to contract smooth muscles. The approach most often employed in the study of angiotensin receptors has been radioactive binding: alteration in the radioactive hormone is detected upon interaction with target cells (Goodfriend and Lin, 1969, 1970; Devynck, Pernollet, Meyer, Fermandjian, and Fromageot, 1973; Glossmann, Baukal, and Catt, 1974).

In our laboratory, in addition to such radioactive-binding experiments, we have used the spin-label monitoring technique to detect alterations in target cells upon interaction with hormone (Schreier-Muccillo, Niculitcheff, Oliveira, Shimuta, and Paiva, 1974). For these studies of peptide hormone-receptor interactions, we have used a preparation of smooth muscle cell membranes from guinea pig intestine.

CHARACTERIZATION OF SMOOTH MUSCLE PLASMA MEMBRANES

Electron Microscopy

Smooth muscle cell membranes from guinea pig intestine were obtained according to the method of Oliveira and Holzhacker (1974).

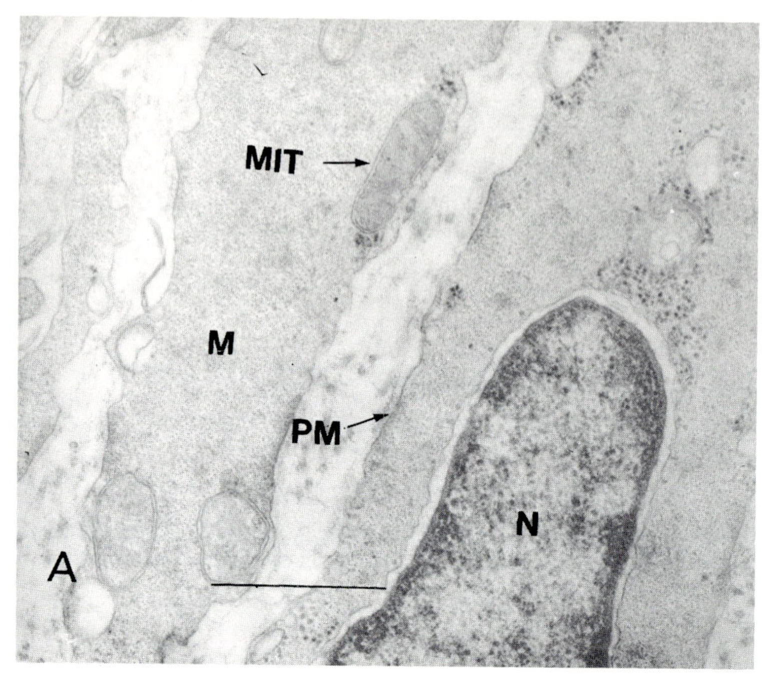
MIT →
M
PM →
N
A

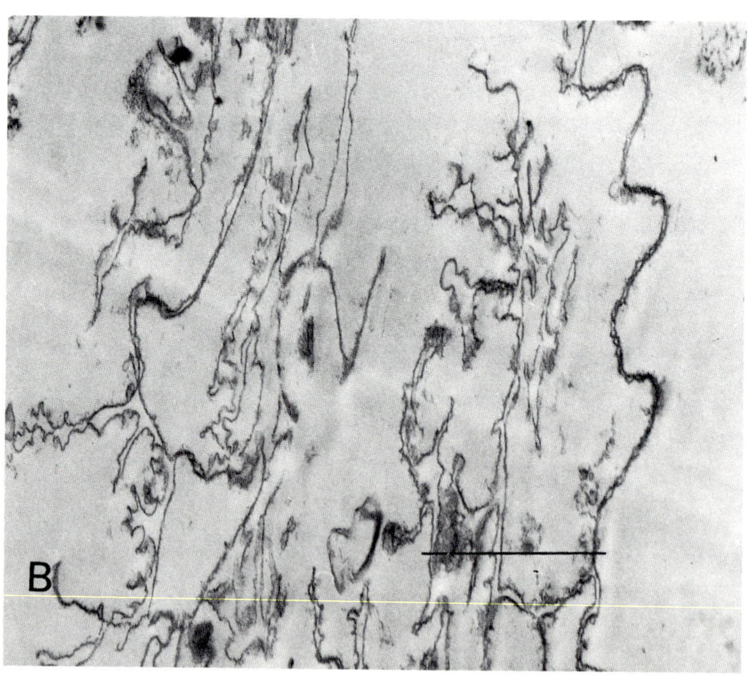
B

The electron microscopic appearance of isolated smooth muscle cell sarcolemma is shown in Fig. 1. At the low-speed centrifugation needed to obtain the cell envelopes there is no apparent vesiculation of the membranous sheets. The membranes thus prepared seemed to be free of contaminating organelles, as confirmed by enzymatic properties.

Chemical Composition

The general chemical composition of smooth muscle cell membranes (Oliveira and Shimuta, 1974) is shown in Table 1. The values obtained for protein, lipid, and carbohydrate are similar to those found for skeletal muscle sarcolemma (Kono, Kakuma, Homma, and Fukuda, 1964; Madeira and Antunes-Madeira, 1973).

TABLE 1. *Chemical composition of smooth muscle cell membranes*

Fraction	% Dry weight	No. of preparations
Protein	62 ± 4	16
Carbohydrate	3 ± 0.2	3
Lipid	20 ± 2	16
	μg/mg Lipid	
Phospholipid	67 ± 5	3
Cholesterol	20 ± 1	3

Data reported are averages of different preparations ± SE.

The high protein:lipid ratio found in both smooth and skeletal muscle sarcolemma (3:1) may be an indicator of specialized functions of these organelles. As pointed out by Guidotti (1972), membranes with enzymatic and transport functions, in addition to the permeability barrier, will have an increased amount of protein.

The phospholipid fraction was considerably lower than that found in striated muscle, showing a difference between these two specialized muscles. The major phospholipid component (42% of total) was phosphatidylcholine (PC), as has been found in a variety of cell membranes (Guidotti,

FIG. 1. Electron micrograph of guinea pig intestinal smooth muscle and isolated sarcolemma. The specimens were fixed in 2% glutaraldehyde, post-fixed in 1% OsO_4 with 0.5% uranyl acetate and embedded in araldite. The ultrathin sections were stained with uranyl acetate and lead acetate. *A:* Intact muscle showing nucleus (N), mitochondria (Mit), myofilaments (M), plasma membrane (PM). *B:* Plasma membrane fraction. Bars represent 1 μm.

1972). Phosphatidylethanolamine (PEA) accounted for 27%, sphingomyelin for 20%, and phosphatidylserine (PS) for 10% of the total phospholipid fraction.

The neutral lipids fraction, separated by thin-layer chromatography, showed among its components: cholesterol esters, free cholesterol, triglycerides, and diglycerides.

Enzymatic Activities

ATP phosphohydrolase [(Na^+, K^+) Mg^{++}-ATPase] (EC 3.6.1.3) is an enzymatic marker for plasma membranes (Skou, 1965; Benedetti and Emmelot, 1968), and it was found in our sarcolemmal fraction.

We have also detected, independent of divalent cations, ATPase that required cations for its maximal activity. $MgCl_2$ enhanced the rate of ATP hydrolysis and the addition of sodium and potassium doubled it. The increment due to Na^+ and K^+ ions could be inhibited by 5 mM ouabain. Addition of $CaCl_2$ (1 mM) had no effect on the Mg^{++}-dependent ATPase (Fig. 2), but in the absence of Mg^{++}, calcium could stimulate the basal activity in the concentration range of 1 μM to 1 mM.

The central role of calcium in the excitation-contraction coupling of skeletal muscle is well established (Sandow, 1965). The mechanism for calcium removal from the cytoplasm in skeletal muscle through the sarcoplasmic reticulum is also well known (Hasselbach, 1964; Inesi, 1972), as is the role of Ca^{++}-activated ATPase in the process.

A homologous calcium-accumulating mechanism has been suggested for smooth muscle, but the site or sites of Ca deposit within the cell are not well defined. An ATP-dependent calcium uptake has been observed in the mitochondrial and microsomal fractions of uterine smooth muscle (Carsten, 1969; Batra and Daniel, 1971), in microsomal fractions of vascular smooth muscle (Fitzpatrick, Landon, Debbas, and Hurwitz, 1972; Baudoin, Meyer, Fermandjian, and Morgat, 1972), and in the plasma membrane fraction of microsomes of guinea pig intestinal smooth muscle (Hurwitz, Fitzpatrick, Debbas, and Landon, 1973).

An ATPase directly activated by 1 mM Ca^{++}, in the absence of magnesium, was observed in plasma membranes of skeletal muscle (Peter, 1970), cardiac muscle (Stam, Shelburne, Feldman, and Sonnenblick, 1969; Dietze and Hepp, 1971), and smooth muscle (Oliveira and Holzhacker, 1974). The concentration of calcium used (1 mM) was higher than would be necessary to activate the ATPase system of the "calcium pump."

Experiments were performed, then, in which the concentration of Ca^{++} was lowered by the use of ethylene glycol bis(β-aminoethyl ether)-N,N'-tetraacetic acid (EGTA) in the mixture. The free calcium concentration was 7.1 μM, as calculated from the value of $3.95.10^{-6}$ M for the dissociation

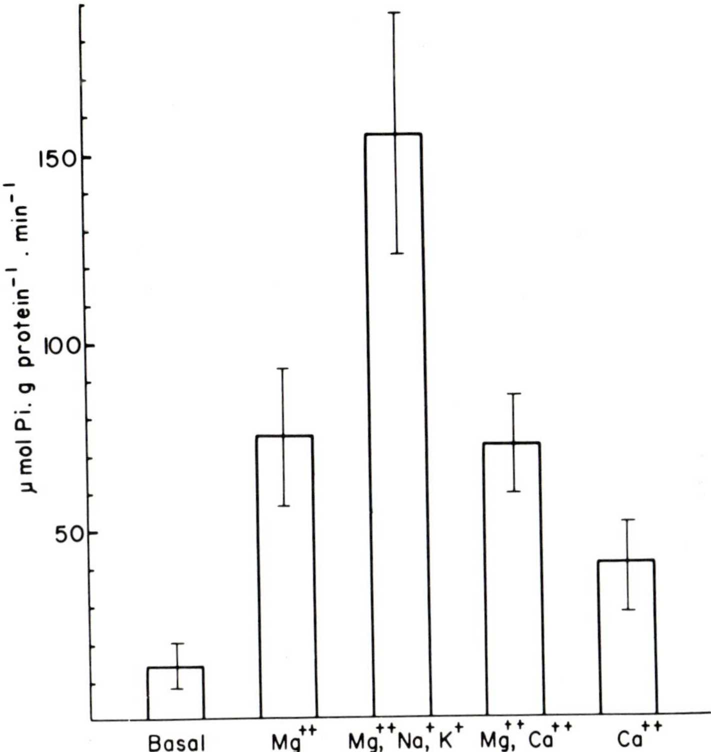

FIG. 2. Effects of cations on ATPase activity of smooth muscle cell membranes. The assay medium consisted of 10 mM Tris-maleate buffer (pH 7.4), 1 mM ATP, and 65–150 μg of membrane protein. The reaction was started with the particulate fraction, incubated for 5 min at 37°C and stopped with 10% trichloroacetic acid. Additions to the medium: Mg^{++}—3 mM $MgCl_2$; Mg^{++}, Na^+, K^+—3 mM $MgCl_2$, 120 mM NaCl, and 12 mM KCl; Mg^{++}, Ca^{++}—3 mM $MgCl_2$, 1 mM $CaCl_2$; Ca^{++}—1 mM $CaCl_2$. The values represent the mean of 7 preparations $\pm$ SE.

constant (de Meis and Hasselbach, 1971) and the conditions of reaction were those reported by de Meis (1971) to be active for the ATPase and calcium uptake in skeletal muscle microsomes. Table 2 shows the values found for Ca^{++}-activated (Mg^{++}-dependent) ATPase in subcellular fractions of smooth muscle cells. The microsomal fraction showed the highest activity, the mitochondrial fraction somewhat less, and none was found in the sarcolemma.

In the plasma membrane fraction, a stimulation (about 20%) of the Mg^{++}-dependent ATPase was observed only when Ca^{++} was lowered to 1 μM (Fig. 3). Higher calcium concentrations failed to produce that effect. However, in the absence of magnesium, the basal ATPase activity was stimulated by calcium in the range of 1 μM to 1 mM.

TABLE 2. *Ca^{++}-activated ATPase in subcellular fractions of smooth muscle cells*

Experiment no.	Fraction	ATPase activity (μmole P$_i \cdot$ mg$^{-1} \cdot$ min^{-1})		
		Mg^{++} dependent	Total: Mg^{++} plus Ca^{++}	Ca^{++} activated
1	Sarcolemma	0.120	0.120	0
	Mitochondria	1.230	1.200	0
	Microsomes	0.870	1.200	0.430
2	Sarcolemma	0.040	0.040	0
	Mitochondria	0.742	0.765	0.023
	Microsomes	0.920	1.040	0.120
3	Sarcolemma	0.065	0.065	0
	Mitochondria	0.412	0.489	0.066
	Microsomes	0.489	0.594	0.105
4	Sarcolemma	0.120	0.120	0
	Mitochondria	0.930	1.000	0.070
	Microsomes	0.770	1.200	0.430

The incubation medium consisted of 10 mM Tris-maleate buffer (pH 7.1), 4 mM MgCl$_2$, 4 mM potassium oxalate, 0.2 mM CaCl$_2$, 0.3 mM EGTA, and 1 mM ATP. For determination of Mg^{++}-dependent ATPase, CaCl$_2$ and potassium oxalate were omitted and the concentration of EGTA was 0.5 mM. The reaction, in a volume of 1 ml, was started with the particulate fraction (100 μg of protein), incubated for 5 min at 37°C, and stopped by Millipore filtration.

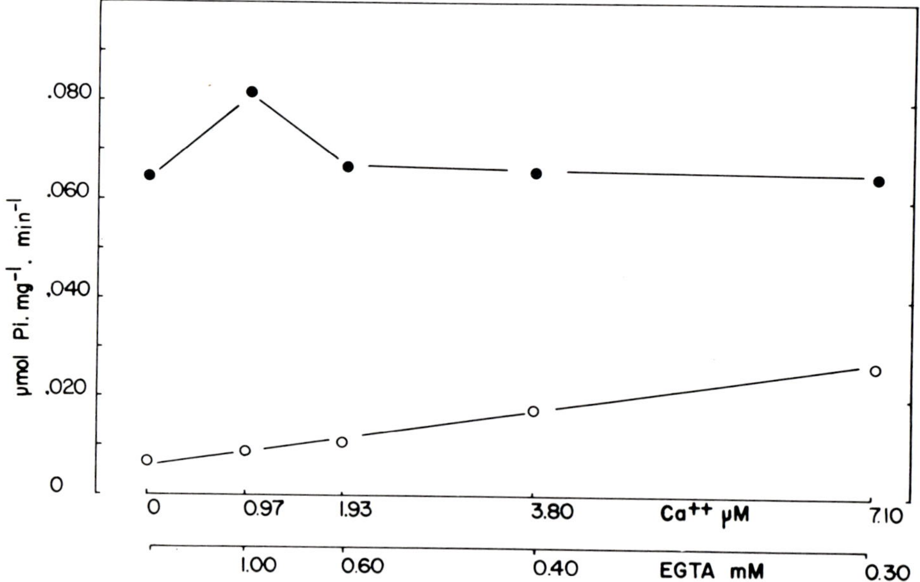

FIG. 3. Effect of low Ca^{++} concentration on the ATPase of smooth muscle sarcolemma. Conditions of reaction are the same as described in Table 2. O———O, without magnesium. ●———●, in the presence of magnesium.

STUDIES ON ANGIOTENSIN RECEPTORS

Binding Experiments

In the study of angiotensin-receptor interaction our first step was to determine the binding of the radioactive hormone with different subcellular fractions of smooth muscle cell. The method used in our experiments was described by Cuatrecasas (1971) for studies of the interaction of insulin with cell membranes of adipose tissue. It consists essentially of a rapid equilibrium of the hormone with target cell membranes, Millipore filtration, and counting of the radioactive material retained in the filter. We employed a tritiated [Asn^1,Val^5]-angiotensin II (gift from Dr. P. A. Khairallah, Cleveland Clinic, Ohio) with a specific activity of 40 Ci/mmole. It showed a single spot in electrophoresis at pH 5.0 and 9.0 and retained full biological activity as assayed in the guinea pig ileum.

Aliquots of freshly prepared particulate fractions (1 to 2 mg protein) were incubated for 10 min at 37°C with ^{3}H-AII (final concentration 10^{-8} M) and diluted to 1 ml with Tyrode solution. Millipore filtration was used to separate bound from free AII. Correction for adsorption of angiotensin to the filters in the absence of cell fractions was effected. "Specific binding" was obtained by subtracting the amount that was not displaced by native 10^{-5} M angiotensin II from the total radioactive uptake.

Figure 4 shows that specific binding of angiotensin to the subcellular fractions of smooth muscle occurred selectively. Plasma membranes bound AII to a larger extent than mitochondria, whereas no uptake of hormone was achieved with microsomes. Figure 4 also shows the enzymatic activities of the three fractions isolated from smooth muscle cell (Oliveira and Holzhacker, 1974), pointing out their relative functional homogeneity. Succinate cytochrome C reductase (a mitochondrial marker) was, in fact, found in higher specific activity in the mitochondrial fraction, whereas no activity was detected in plasma membranes and little in the microsomal fraction. The differential distribution of glucose-6-phosphatase also followed the same general scheme, the microsomal fraction being the one with the highest activity, as would be expected. The presence of Na$^+$,K$^+$-ATPase in the plasma membrane fraction was an indication of the functional state of the membrane. Our preliminary results with radioactive angiotensin show a higher affinity of the hormone for plasma membranes than for the other subcellular fractions of the intestinal smooth muscle. This binding, observed with the hormone concentration needed for half maximum contraction in the intact ileum, could be displaced by native angiotensin.

Although the earlier experiments of Goodfriend and Lin (1970) showed binding of angiotensin in all three subcellular fractions of target tissues, more recent work shows angiotensin binding specifically to plasma membranes of rabbit aorta smooth muscle (Devynck et al., 1973) and plasma

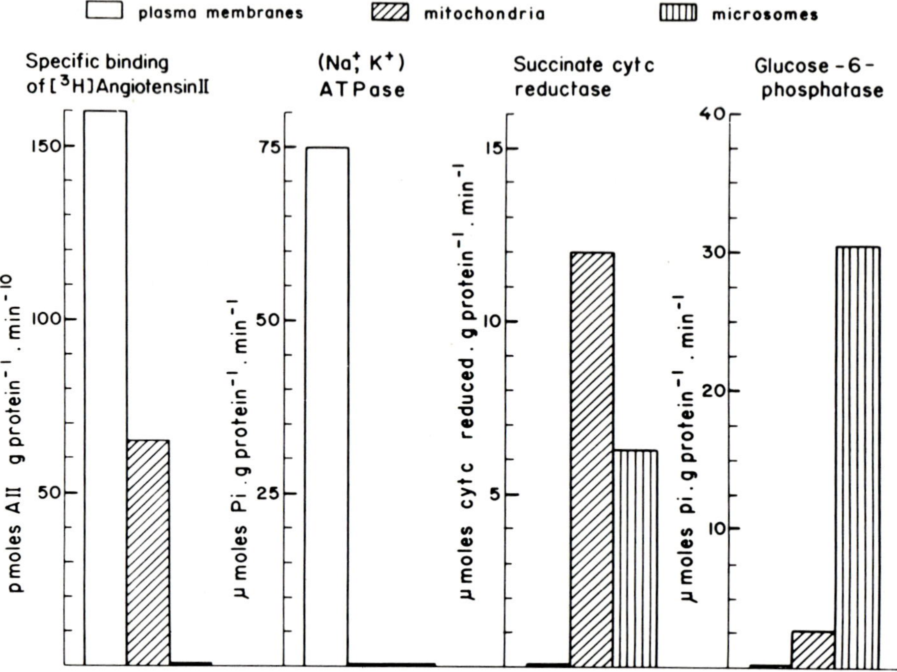

FIG. 4. Binding of ³H-AII and enzymatic activities of subcellular fractions of smooth muscle cells. Details of binding experiments in the text. Enzymatic activities from data of Oliveira and Holzhacker (1974).

membranes of bovine and rat adrenal glands (Glossmann et al., 1974). These findings corroborate the current view that polypeptide hormones interact first with the plasma membranes of target cells (Rasmussen, 1971). Binding may be a first step, leading to conformational changes at the membrane as part of the mechanism of hormone action.

Spin-label Experiments

We have studied the possible conformational changes taking place at the membranes upon interaction with AII, using the spin-label technique, which has been proved useful for studies of structure and function of biological membranes (McConnell and McFarland, 1970; Schreier-Muccillo and Smith, 1975).

We have studied AII interaction with plasma membranes from smooth muscle as well as with model systems. Only the former underwent changes in the presence of this hormone.

Plasma Membranes

Protein label. Membranes (0.8 to 1.5 mg protein/ml) were suspended in Tyrode solution. The incorporation of protein label was performed at 4°C, with 0.010 to 0.013 mg label/mg membrane protein. The incubation period was usually 12 hr. Membranes containing protein-specific labels displayed spectra with two components — one characteristic of label bound to sites of higher mobility, and another indicative of sites where the motion of the label was greatly restricted (McConnell and McFarland, 1970). Figure 5 shows the spectra of 4-maleimido-2,2,6,6-tetramethyl-piperidinooxyl (MSL-6) in the membranes, and the effect of addition of 1 mM solution of angiotensin. The hormone produced an increase in the population of labels in a more hindered environment (Fig. 5, *dotted line*).

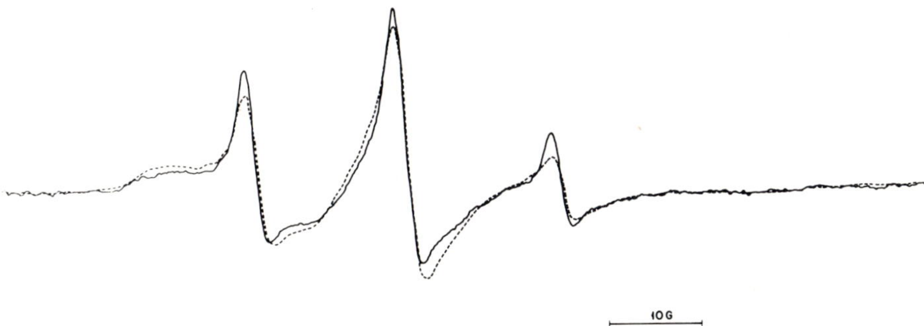

FIG. 5. Spectra of MSL-6 in plasma membranes from smooth muscle suspended in Tyrode solution. *Solid line:* in the absence of AII. *Dotted line:* in the presence of 1 mM solution of AII. G stands for Gauss.

Lipid label. The lipid spin probe, at ratios of 0.003 to 0.0125 mg/mg membrane protein, were intercalated by uptake from a film of probe deposited in a test tube, and the spectra taken after 3 to 4 washings. Only 2-(3-carboxypropyl)-4,4-dimethyl-2-tridecyl-3-oxazolidinyloxyl (5-SASL) was used for membranes. Some preparations with higher lipid probe concentration displayed some exchange-broadened spectra. Samples were placed in Pasteur pipettes.

Spectra of the stearic acid probe in aqueous suspensions of membranes also indicated two different states (Fig. 6) — one of a high degree of mobility, and another in which motion was hindered (membrane-bound probes). The narrow lines have already been observed in other membrane preparations (Landsberger, Lenard, Paxton, and Compans, 1971; Seelig and Hasselbach, 1971; Landsberger, Compans, Paxton, and Lenard, 1972; Kaplan, Canonico, and Caspary, 1973). They seem to be due to probes

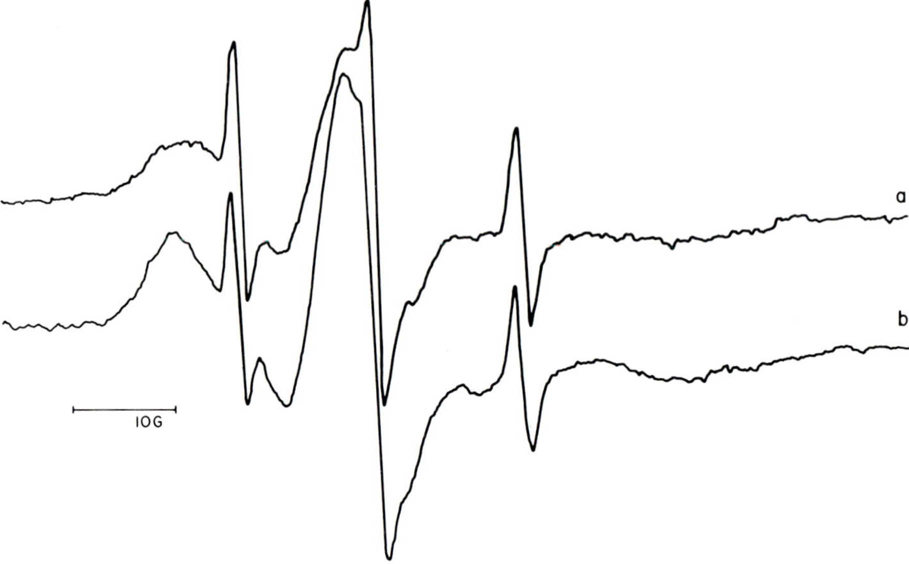

FIG. 6. Spectra of 5-SASL in plasma membranes from smooth muscle suspended in Tyrode solution: *a*, in the absence of AII; *b*, in the presence of a 1 mM solution of AII. G stands for Gauss.

tumbling more freely in an aqueous environment, as measured by the hyperfine splittings and line widths (Butler, Tattrie, and Smith, 1974). A decrease in their intensity occurred in presence of the hormone. Addition of AII (final concentrations 1 mM) to an aqueous solution of 5-SASL caused no spectral changes.

The effect of angiotensin was concentration dependent. The empirical parameter used in Table 3 shows large effects at 1 mM angiotensin, which

TABLE 3. *Effect of angiotensin II and analogue peptides on the spectra of 5-SASL-labeled membranes*

Experiment[a]	h_m/h_f[b] with no peptide added	Peptide added	Concentration (mM)	h_m/h_f[b] after addition of peptide	% change
1	0.420	AII	1.00	0.840	100
2	0.437	AII	1.00	0.840	92
3	0.228	AII	0.10	0.375	65
4	0.409	AII	0.10	0.623	52
5	0.256	AII	0.01	0.281	10
6	0.344	Leu[8]-AII	1.00	0.448	30
7	0.465	Gly[1], Pro[8]-AII	1.00	0.598	29

[a] The experiments described in this Table are representative of a large number of observations.

[b] h_m/h_f is the ratio of the heights of the low-field lines, corresponding to membrane-bound (h_m) and freely tumbling (h_f) spin probes.

decreased when 0.1 mM AII was added, and were almost negligible for 0.01 mM AII. The changes were smaller for 1 mM AII when such concentration was added to preparations previously treated with 0.01 mM hormone.

The alterations promoted by two analogues: [Leu^8]-AII (intrinsic activity 48% that of AII, and $pD_2 = 5.00$, as compared with $pD_2 = 8.80$ for AII) (Paiva, Goissis, Juliano, Miyamoto, and Paiva, 1974) and [Gly^1, Pro^8]-AII (intrinsic activity 96%, $pD_2 = 4.00$) were qualitatively similar to those observed with AII, but occurred to a lesser extent (Table 3). pD_2 is the negative logarithm of the drug-receptor dissociation constant, taken as the concentration needed to produce half-maximum response.

Delipidated membranes retained some of the lipid probe, yielding spectra which also presented the narrow lines. The relative intensity of the latter also decreased upon addition of AII.

Model Systems

Lipid dispersions and planar multibilayers were prepared as described (Schreier-Muccillo, Marsh, Dugas, Schneider, and Smith, 1973). The spectra yielded by dispersions (containing 5-SASL) and films containing 17β-hydroxy-4′,4′-dimethyl spiro (5α-androstane-3,2′-oxazolidin)-3′-yloxyl (ASL) of PC, PC-cholesterol, PC-PS, PC-PS-cholesterol, and PC-PA-cholesterol underwent no changes when AII was added. The same was true for aqueous dispersions of membrane lipids (with 5-SASL).

The changes in the spectra of spin labels covalently bound to plasma membrane proteins indicate that the hormone caused a decreased mobility of a number of sites. These results are in line with the findings of Oliveira and Holzhacker (1974) that the hormone caused contraction of the isolated plasma membranes. However, the diminished intensity of the narrow lines in the spectra of 5-SASL upon addition of the hormone suggests an enhancement of the fluidity of membrane lipid regions, and, as a consequence, an increased "solubility" of spin probes in the membrane. We ascribe this event to the small lipid:protein ratio (1:3) of our membranes. It is conceivable that the membrane lipids would tend to expand in order to accommodate protein contraction, but since there is a large proportion of protein, the overall result is the observed contraction.

It should be noted that delipidated membranes gave the same results as intact membranes. This could be explained by the presence of some lipid not removed during the delipidation procedure which would display the same behavior as the total lipid. However, it cannot be ruled out that the relative decrease in the intensity of the lines due to more freely tumbling 5-SASL might be caused by a tight binding of the probe to membrane proteins as a result of the interaction with angiotensin.

That the site of action of the hormone must be protein in nature was shown by the absence of an effect in systems containing only lipids.

Although other authors have detected conformational alterations at membranes at lower concentrations of other hormones (Rubin, Swislocki, and Sonnenberg, 1973; Postel-Vinay, Sonnenberg, and Swislocki, 1974; Kury, Ramwell, and McConnell, 1974), in our experiments the concentrations needed for the observation of spectral changes were considerably higher when compared to pharmacological conditions. One possible explanation for the high-AII concentrations needed to produce the changes reported here may be that a partial loss of the effector system occurred during the preparation of membranes. This may mean that nonspecific binding was also monitored by our experiments. But we believe that hormone-receptor interaction was also detected, in view of the result obtained when 1 mM AII was added to membranes after 0.01 mM hormone had been previously added. This experiment resembles the observation of tachyphylaxis by Paiva, Juliano, Nouailhetas, and Paiva (1974) and suggests that some of the effects revealed by the spin labels reflect the specific interaction. The smaller effect of the analogues is also suggestive of specificity. It should also be taken into account that although the threshold concentration for activity of angiotensin in the isolated guinea pig ileum is 10^{-9} M, that may not be the concentration at the membranes of target cells.

As for the mechanism of action of angiotensin, our results indicate that when the hormone binds to the membrane, it causes a series of conformational changes, as evinced by protein-bound spin labels and by lipid spin probes. Whether these events are a first step in the mechanism of hormone action, or a consequence of other events occurring prior to them has still to be clarified.

ACKNOWLEDGMENTS

We thank Mr. J. Alves for help with the experimental work. The research reported here was supported by grants from the São Paulo State Research Foundation (FAPESP) and the Brazilian National Research Council (CNPq).

REFERENCES

Batra, S. C., and Daniel, E. G. (1971): ATP-dependent Ca^{++} uptake by subcellular fractions of uterine smooth muscle. *Comp. Biochem. Physiol.* 38A:369–385.

Baudoin, M., Meyer, P., Fermandjian, S., and Morgat, J. L. (1972): Calcium release induced by interaction of angiotensin with its receptors in smooth muscle cell microsomes. *Nature* 235:336–338.

Benedetti, E. L., and Emmelot, P. (1968): Structure and function of plasma membranes isolated from liver. In: *The Membranes,* edited by A. J. Dalton and F. Haguenau, pp. 33–120. Academic Press, New York.

Butler, K. W., Tattrie, N. H., and Smith, I. C. P. (1974): The location of spin probes in two phase mixed lipid systems. *Biochim. Biophys. Acta,* 363:351–360.

Carsten, M. E. (1969): Role of calcium binding by sarcoplasmic reticulum in the contraction and relaxation of uterine smooth muscle. *J. Gen. Physiol.* 53:414–426.

Cuatrecasas, P. (1971): Insulin-receptor interactions in adipose tissue cells: direct measurement and properties. *Proc. Nat. Acad. Sci. U.S.A.* 68:1264–1268.

de Meis, L. (1971): Allosteric inhibition by alkali ions of the Ca^{++} uptake and adenosine triphosphatase activity of skeletal muscle microsomes. *J. Biol. Chem.* 246:4764–4773.

de Meis, L., and Hasselbach, W. (1971): Acetyl phosphate as substrate for calcium uptake in skeletal muscle microsomes. *J. Biol. Chem.* 246:4759–4763.

Devynck, M. A., Pernollet, M. G., Meyer, P., Fermandjian, S., and Fromageot, P. (1973): Angiotensin receptors in smooth muscle cell membranes. *Nature New Biol.* 245:55–57.

Dietze, G., and Hepp, K. D. (1971): Calcium stimulated ATPase of cardiac sarcolemma. *Biochem. Biophys. Res. Comm.* 44:1041–1049.

Fitzpatrick, D. F., Landon, E. J., Debbas, G., and Hurwitz, L. (1972): A calcium pump in vascular smooth muscle. *Science* 176:305–306.

Glossmann, H., Baukal, A. J., and Catt, K. J. (1974): Properties of angiotensin II receptors in the bovine and rat adrenal cortex. *J. Biol. Chem.* 249:825–834.

Goodfriend, T., and Lin, S. Y. (1969): Angiotensin receptors. *Clin. Res.* 17:243.

Goodfriend, T., and Lin, S. Y. (1970): Receptors for angiotensin I and II. *Circ. Res.* (Suppl. I) 26:163–174.

Guidotti, G. (1972): The composition of biological membranes. *Arch. Int. Med.* 129:194–201.

Hasselbach, W. (1964): Relaxing factor and relaxation of muscle. *Progr. Biophys.* 14:169–222.

Hurwitz, L., Fitzpatrick, D. F., Debbas, G., and Landon, E. R. (1973): Localization of calcium pump activity in smooth muscle. *Science* 179:384–386.

Inesi, G. (1972): Active transport of calcium ion in sarcoplasmic membranes. *Annu. Rev. Biophys. Bioeng.* 1:191–210.

Kaplan, J., Canonico, P. G., and Caspary, W. J. (1973): Electron spin resonance studies of spin-labeled mammalian cells by detection of surface-membrane signals. *Proc. Nat. Acad. Sci. U.S.A.*, 70:66–70.

Kono, T., Kakuma, F., Homma, M., and Fukuda, S. (1964): The electron-microscopic structure and chemical composition of the isolated sarcolemma of the rat skeletal muscle cell. *Biochim. Biophys. Acta* 88:155–176.

Kury, P. G., Ramwell, P. W., and McConnell, H. M. (1974): The effect of prostaglandins E_1 and E_2 on the human erythrocyte as monitored by spin labels. *Biochem. Biophys. Res. Commun.* 56:478–483.

Landsberger, F. R., Lenard, J., Paxton, J., and Compans, R. W. (1971): Spin label electron spin resonance study of the lipid-containing membrane of influenza virus. *Proc. Nat. Acad. Sci. U.S.A.*, 68:2579–2583.

Landsberger, F. R., Compans, R. W., Paxton, J., and Lenard, J. (1972): Structure of the lipid phase of Rauscher murine leukemia virus. *J. Supramol. Struct.* 1:50–54.

Madeira, V. M. C., and Antunes-Madeira, M. C. (1973): Chemical composition of sarcolemma isolated from rabbit skeletal muscle. *Biochim. Biophys. Acta* 298:230–238.

McConnell, H. M., and McFarland, B. G. (1970): Physics and chemistry of spin labels. *Quart. Rev. Biophys.* 3:91–136.

Oliveira, M. M., and Holzhacker, S. (1974): Isolation and characterization of smooth muscle cell membranes. *Biochim. Biophys. Acta* 332:221–232.

Oliveira, M. M., and Shimuta, S. (1974): Chemical composition of smooth muscle cell sarcolemma. *Ms. in preparation.*

Paiva, T. B., Goissis, G., Juliano, L., Miyamoto, M. E., and Paiva, A. C. M. (1974): Angiotensin-like and antagonistic activities of N-terminal modified |8-leucine| angiotensin II peptides. *J. Med. Chem.* 17:238–241.

Paiva, T. B., Juliano, L., Nouailhetas, V. L. A., and Paiva, A. C. M. (1974): The effect of pH on tachyphylaxis to angiotensin peptides in the isolated guinea pig ileum and rat uterus. *Eur. J. Pharm.* 25:191–196.

Peter, J. B. (1970): A $(Na^+ + K^+)$ ATPase of sarcolemma from skeletal muscle. *Biochem. Biophys. Res. Commun.* 40:1362–1367.

Postel-Vinay, M. C., Sonenberg, M., and Swislocki, N. I. (1974): Effect of bovine growth hormone on rat liver plasma membranes as studied by circular dichroism and fluorescence using the extrinsic probe 7,12-dimethyl-benzanthracene. *Biochim. Biophys. Acta* 332:156–165.

Rasmussen, H. (1971): Cell receptor sites for peptide hormones. *Excerpta Medica* 1:194–198.

Rubin, M. S., Swislocki, N. I., and Sonenberg, M. (1973): Alteration of liver plasma membrane protein conformation by bovine growth hormone "in vitro". *Arch. Biochem. Biophys.* 157:252–259.

Sandow, A. (1965): Excitation-contraction coupling in skeletal muscle. *Pharmacol. Rev.* 17: 265–320.

Schreier-Muccillo, S., Marsh, D., Dugas, H., Schneider, S., and Smith, I. C. P. (1973): A spin probe study of the influence of cholesterol on the smotion and orientation of phospholipids in oriented multibilayers and vesicles. *Chem. Phys. Lipids* 10:11–17.

Schreier-Muccillo, S., Niculitcheff, G. X., Oliveira, M. M., Shimuta, S., and Paiva, A. C. M. (1974): Conformational changes at membranes of target cells induced by the peptide hormone angiotensin. A spin label study. *FEBS Lett.,* 47:193–196.

Schreier-Muccillo, S., and Smith, I. C. P. (1975): Spin labels as probes of biological membranes. Academic Press, New York (*in press*).

Seelig, J., and Hasselbach, W. (1971): A spin label study of sarcoplasmic vesicles. *Eur. J. Biochem.* 21:17–21.

Skou, V. C. (1965): Enzymatic basis for active transport of Na^+ and K^+ across cell membrane. *Physiol. Rev.* 45:596–617.

Stam, A. C., Shelburne, J. W., Feldman, D., and Sonnenblick, E. H. (1969): A miocardial sarcolemma preparation and the ouabain-sensitive $(Na^+ - K^+)$-ATPase. *Biochim. Biophys. Acta* 189:304–307.

Concepts of Membranes in Regulation and Excitation,
edited by M. Rocha e Silva and G. Suarez-Kurtz.
Raven Press, New York © 1975.

A Thermodynamic Approach to Problems of Drug Antagonism: the Microphysical Model

M. Rocha e Silva

Department of Pharmacology, Faculty of Medicine of Ribeirão Preto, University of São Paulo, 14100 Ribeirão Preto, S.P., Brazil

INTRODUCTION

In 1970, we proposed a microphysical model for the phenomenon of recovery of a biological preparation (guinea pig ileum) from inhibition by a fully reversible competitive antagonist to histamine, such as diphenhydramine (Benadryl®). It has long been known (Rocha e Silva and Beraldo, 1948; Beraldo and Rocha e Silva, 1949) that if a strip of guinea pig ileum is exposed to a large concentration of an antihistamine (diphenhydramine) and is then washed out, the recovery process follows the course of an S-shaped line. If at any moment the preparation is tested with a large dose or concentration of the agonist (histamine), it will give a maximum response, but if the agonist is washed out, the muscle comes back to the same previous level of inhibition. This was the basis of a new approach to the phenomenon—the assumption that the antagonist remains in place, bound to an "annex part" of the receptor, while allowing free competition with the "specific part" of the receptor. This phenomenon was called the "Charnière effect" (Rocha e Silva, 1969). Since the antagonist has no access to the annex part of the receptor, the antagonist cannot be displaced by any concentration of the agonist from that part of the receptor site. Further studies strengthened the thermodynamic basis of the Charnière effect: although temporarily or permanently blocked by the presence of the antagonist (Benadryl® or Dibenzyline®), the receptor recovers full reactivity if the material is intermittently exposed to low temperatures (3° to 5°C) and the degree of inhibition is measured at 37°C (Rocha e Silva, Fernandes, and Antonio, 1972; Rocha e Silva and Fernandes, 1974).

THE CHARNIÈRE EFFECT

The idea of an annex part that holds the heavy lipophilic or hydrophobic rings or groups of the antagonist was prompted by the work of Ariëns (1964) and Ariëns and Simonis (1967). An interesting biochemical analogy was described by Baker (1962, 1964) as an "exoalkylation" by covalent binding of inhibitors with chemical groups not necessarily belonging to the active site of the enzyme. Another example of a Charnière effect can be found in the beautiful experiments of Ritchie and Greengard (1966) in which they showed that local anesthetics applied directly to the unsheathed nerve, though firmly bound by the hydrophobic part of the molecule, nonetheless allow a reversible blockade of the nerve transmission by intermittently changing the pH from 7.2 to 9.2 and back to 7.2. More recently, Stewart (1974) described a Charnière effect obtained with the small peptide Pro-Phe-Arg attached to the alkylating nitrogen mustard chlorambucil, which produces a permanent inhibition of the response of the ileum to angiotensin II (Ang II), while allowing full activity of the agonist when added in higher concentrations.

Figure 1(I) pictures a receptor with the two sites—annex (A) and specific (S)—and the two interacting compounds, the antagonist and the agonist. In Fig. 1(II) the constant of affinity (K_n) of the agonist toward the specific part of the receptor and the associated free-energy function $(-\Delta G_a)$ are indicated. In Fig. 1(III) the interaction of the antagonist with the receptor, with an overall affinity K_i and free-energy function $-\Delta G_i$, is shown; it can be seen that the antagonist is bound to the annex part of the receptor by heavy hydrophobic rings, while allowing a competition with the agonist for the specific part of the receptor, as shown in Fig. 1(IV).

The three constants of affinity—K_n, K_n', and K_i—are related through the equation:

$$\beta = \frac{K_n'}{K_n} = 1 + \frac{(I)}{K_i}$$

where K_n is the slope of the double-reciprocal plot $(1/y$ versus $1/x)$ in the absence of the antagonist; K_n' represents the slope of the line in presence of a concentration (I) of the antagonist; β is the ratio of the slopes in presence/absence of the antagonist, and K_i is the so-called overall constant of affinity of the antagonist toward the receptor site. The logarithmic form of this equation is

$$\log (\beta - 1) = \log I - \log K_i$$

Since $-\log K_i = pA_2$, then

$$pA_2 = \log (\beta - 1) - \log I$$

We can then calculate the pA_2 for any value of the concentration (I) of the antagonist in equilibrium with the preparation. The value of $\beta = K_n'/K_n$ can be deduced directly from the slopes of the experimental lines obtained by plotting the reciprocals $(1/y)$ of the effects against the reciprocals of the concentrations $(1/x)$ of the agonist, as shown in Fig. 2.

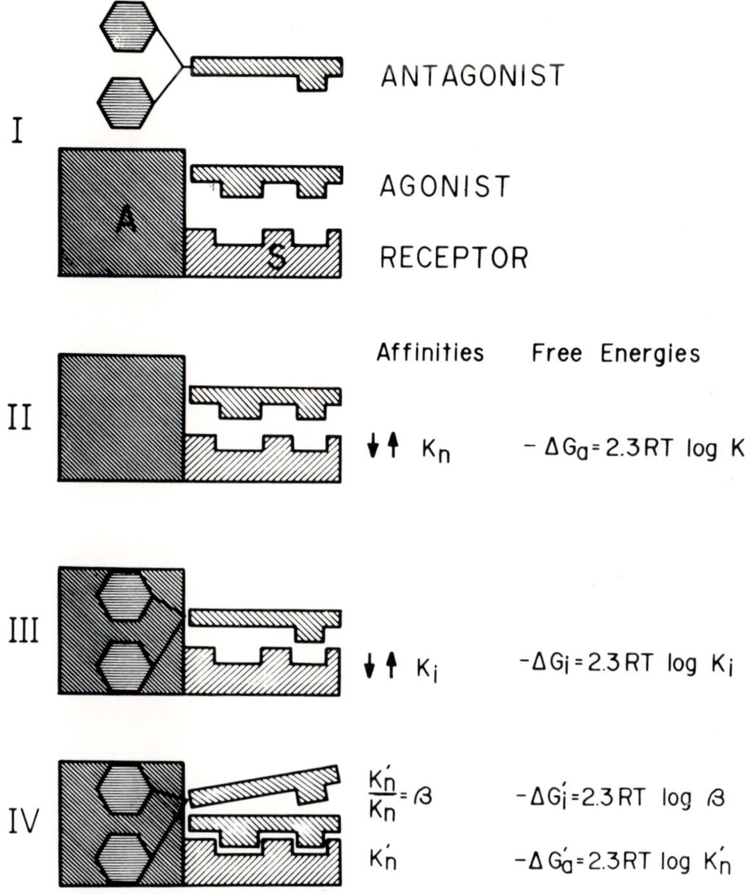

FIG. 1. The Charnière effect. I: Schematic view of the receptor site with an annex (A) and a specific (S) part; the molecules of the agonist and antagonist are represented by their salient interacting groups. II: Interaction of the agonist with the specific part of the receptor; K_n, constant of affinity, and $-\Delta G_a$, free energy of interaction of the agonist with the receptor. III: Idealized conditions in which the antagonist interacts with the receptor site with an overall constant of affinity K_i and free energy $-\Delta G_i$. IV: Competition of the agonist and antagonist at the specific part of the receptor; K'_n is the new affinity constant of the agonist in presence of the antagonist with a free energy function $-\Delta G'_a$; the ratio $\beta = \dfrac{K'_n}{K_n}$ permits to calculate the free energy function $(-\Delta G'_i)$ of the antagonist toward the specific part of the receptor > as the difference $-(\Delta G'_a - \Delta G_a) = 2.3\,RT\,\log \beta$.

The energetic basis for the Charnière effect can be further found by analysis of the parallel lines in Fig. 3. The slope of all lines in this diagram equals 2.3 *RT*, where $R = 1.987$ kcal/molecule/degree, the gas constant, and *T*, the absolute temperature in degrees Kelvin. If we start with an energetic level of $-\Delta G_a$ (the free energy of the agonist in the absence of the

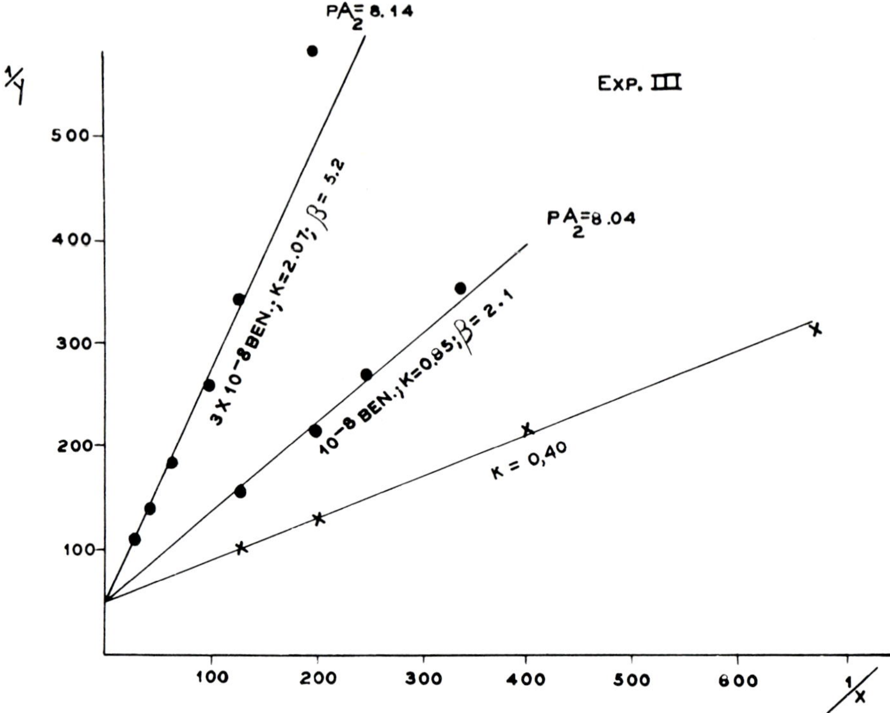

FIG. 2. Double-reciprocal plots for histamine on the guinea pig ileum. *Lower line:* $K = 0.40$, control without antagonist. *Middle line:* Obtained in presence of 10^{-8} diphenhydramine (Ben) with a slope $K' = 0.85$ and a $\beta = K'/K = 2.1$. *Upper line:* Obtained in presence of 3×10^{-8}. Ben with a slope $K' = 2.0$ and a ratio $\beta = 5.2$. The values $pA_2 = 8.04$ and 8.14 were obtained by the equation indicated in the text.

antagonist) and increase the concentration of the antagonist on a log scale, taking K_i ($= I_{pA_2}$) corresponding to the pA_2 level as the unit of concentration of the antagonist, $\log n$ will represent the log concentration of the antagonist, starting from $\log 1 = 0$. In the diagram $-\Delta G'_a$ represents the decrease in free energy (or affinity) of the agonist to the receptor in the presence of increasing concentrations of the antagonist ($n = \beta - 1$) and linearly related to $\log n$; the difference $\Delta G'_a - \Delta G_a = \Delta G'_i$ indicates the extra free energy needed to displace the antagonist from the specific part of the receptor: $-\Delta G'_i = 2.3 \, RT \log \beta$, as shown in Fig. 1(IV). The upper lines in Fig. 3 refer to the overall interaction of the antagonist with the receptor site, where $-\Delta G_i = 2.3 \, RT \log K_i = 2.3 \, RT \log pA_2$ indicates the variation of free energy of the interaction of the antagonist with the receptor, at the level of the pA_2, and $-\Delta G_i^n = 2.3 \, RT \log nK_i$ represents the so-called "chemical potential" for increasing concentrations of the antagonist, its

PARAMETERS OF AFFINITY

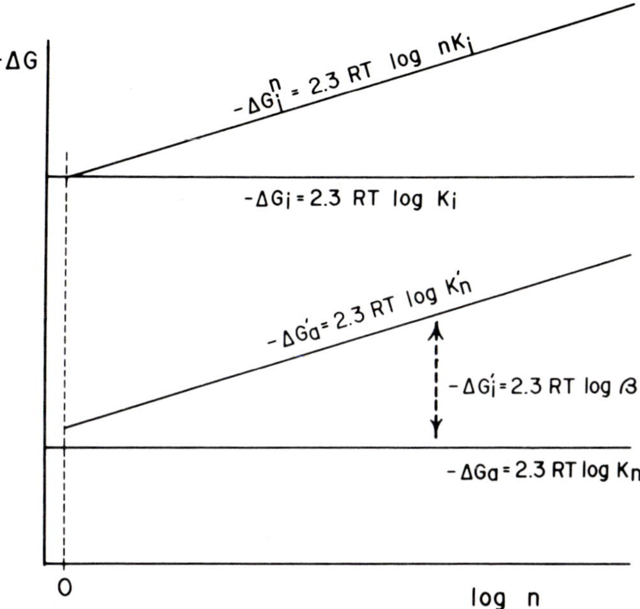

FIG. 3. A thermodynamic analysis of the Charnière effect. In the abscissae, increasing concentrations of the antagonist in units of concentration K_i ($C = nK_i$); in the ordinates, the values of free energies ($-\Delta G$) of interaction of the agonist and antagonist with the receptor site.

initial value coinciding with the standard free energy ($-\Delta G_i$) of the antagonist when $n = 0$ òr $I = K_i$.

Another useful concept is that formulated more recently (Rocha e Silva and Fernandes, 1974) of "pA$_2$ at the level of the receptor," a situation that prevails when the unbound antagonist has been washed away from the outside bath. With a typical synthetic antihistaminic such as diphenhydramine (Benadryl®), when the preparation is washed with new Tyrode solution, there is a progressive tendency of the pA$_2$ to return to 0, and of the preparation to return to its full basic sensitivity towards the agonist. With phenoxybenzamine (Dibenzyline®), however, after a contact of the antagonist with the preparation, the antagonism persists at a stabilized level (pA$_2$ = 9.00) for several hours, with continuous washings at 37°C. From the persistence of the inhibition after repeated washings with new Tyrode, one can conclude that the antagonist is firmly held at the receptor site and therefore the calculated values of pA$_2$ under those conditions could measure the apparent constant of dissociation of the antagonist ($-\log K_i$) at the level of the receptor site. On the basis of those findings, it was possible to study the effect of temperature upon the formation of the bindings of Dibenzyline® with the receptor for histamine and deduce the values of enthalpy and entropy of the bindings by purely pharmacological means (Rocha e Silva and Fernandes, 1974).

THE MICROPHYSICAL MODEL

On this thermodynamical basis we have proposed a microphysical model to account for the recovery phenomenon, assuming that the heavy lipophilic rings of the antagonist are held by hydrophobic and van der Waals forces to the annex part of the receptor, as indicated in Fig. 1(III). That intermittent exposure to low temperatures (from 3° to 5°C) will disrupt the bindings of the antagonist to this part of the receptor gives strong support to the lipophilic or hydrophobic character of such linkages. To account for the S-shaped curve of the recovery process, we have proposed in a previous publication (Rocha e Silva, 1970) a double-exponential equation:

$$N_i = N_o e^{-\Delta A e^{-kt}} \tag{1}$$

where N_o is the number of molecules of the antagonist vibrating inside a potential well, the energy of which can be taken as

$$-\Delta G_i = 2.3 \; RT \; pA_2 \tag{2}$$

Figure 4 gives a schematic view of the potential well with depth ΔG_i and maximum excursion $\Delta x \simeq 10^{-8}$ cm. The equation presented in the picture is the one given by Steinberg and Sheraga (1963) to calculate the frequency of a molecule of weight M, vibrating in a potential well. N_i is the number of molecules of average energy ΔA that will overcome the potential barrier

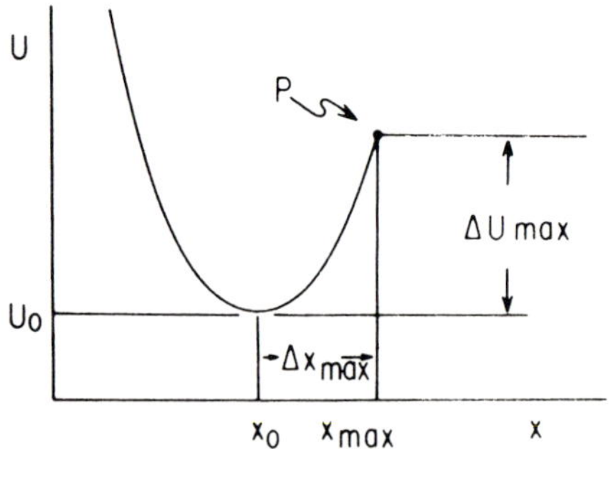

$$Y = \frac{1}{\pi \Delta x_{max}} \sqrt{\frac{\Delta U_{max}}{M}}$$

FIG. 4. Schematic view of the potential well of depth ΔU_{max} where the molecule of the antagonist is held by hydrophobic bindings to the annex part of the receptor. The equation indicated was deduced by Steinberg and Scheraga (1963) to calculate the frequency of vibration of a molecule of weight M in that potential well. To complete the calculation of the free energy of vibration, see Table I.

at time t (in minutes); k is the time constant, the reciprocal of which $(1/k)$ will give the time in which the number of molecules escaping the well will be

$$N_i = N_o e^{-0.37\Delta A} \tag{3}$$

If we assume that ΔA decreases exponentially with time, and make ΔA_o the initial value of ΔA at time 0, then the exponential equation

$$\Delta A = \Delta A_o e^{-kt} \tag{4}$$

is incorporated in the strength of the double-exponential equation given above (Eq. 1). A good agreement of the experimental data with the calculated values is indicated in Fig. 5.

Now to complete the model we have to find out which energy value best represents ΔA in Eq. 4. We must consider two obvious possibilities; first that $-\Delta G_i$ is the free energy binding the antagonist to the receptor site. However, according to Eqs. 1 and 4 this value would constitute an over-estimation of the energy, which might throw the molecules out of the potential well, because part of the energy is counterbalanced by the vibration energy, as calculated in Table I. The second possibility would be to consider the vibration energy ($\mu_{\text{vibr.}}$) as the force pushing the molecules out of the well. This choice would also be unfortunate because the energy calculated by quantum mechanics is fairly constant for a wide range of values of $-\Delta G_i$, i.e., it could be an overestimation up to a certain pA_2 value and an underestimation as the affinity of the antagonist to the receptor site in-

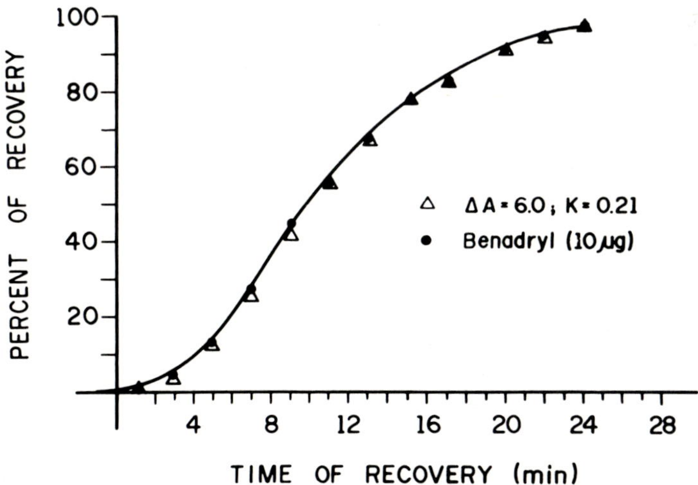

FIG. 5. Curves of recovery from inhibition by diphenhydramine (10 μg of Benadryl®) (●——●) and the theoretical curve (Δ—Δ) using Eq. 1 when $A = \Delta A = 6.0$ and $K = 0.21$. The antagonist has been kept for 1 min in contact with the preparation and then washed away. Every 2 min the preparation is tested with 0.1 μg histamine.

TABLE 1. *Calculation of free energies of vibration ($\mu_{\text{vibr.}}$) corrected for degrees of freedom ($\approx$ 120) and heat capacity (3 RT).*

pA$_2$	ΔG_i (Kcal/mole)	Basic mode of vibration[a]: $\mu^o_{\text{vibr.}}$	Corrected for degrees of freedom[b]: $\mu_{\text{vibr.}}$	Total[c]: $\mu_{\text{vibr.}} + 3\,RT$	$\Delta A = \Delta G_i - $ total[d]
2	2.852	1,440	4,408	6.268	−3.416
3	4.278	1,315	4,283	6.143	−1.865
4	5.704	1,226	4,194	6.054	−0.350
5	7.130	1,168	4,136	5.996	+1.134
6	8.566	1,157	4,125	5.985	+2.581
7	9.982	1,052	4,020	5.880	+4.102
8	11.406	1,011	3,979	5.839	+5.567
9	12.834	974.5	3,942	5.802	+7.032
10	14.260	941.9	3,910	5.770	+8.489

[a] The basic vibration mode in terms of Helmholtz free energy is $\mu^o_{\text{vibr.}} = -RT \ln Z$, where Z is the partition function that can be approximated to $kT/h\nu$. To calculate the frequency given by the formula in Fig. 6 we take $\Delta U_{max} = \Delta G_i$ (in ergs) $= 2.3\,RT\,\text{pA}_2\,(\times\,4.18 \times 10^{-7})$; we can consider the maximum excursion $\Delta x = 10^{-8}$, i.e., of the order of the Ångstrom; $k = 1.38 \times 10^{-16}$ erg/degree (Boltzmann constant); $h = 6.628 \times 10^{-27}$ erg/sec (Planck's constant); $T = 310°K$ (absolute temperature), and $R = 1.987$ Kcal/mole/degree (gas constant).

[b] We take the average value of 120 as the number of degrees of freedom of vibration of a molecule such as diphenhydramine and calculate a correction parcel $-RT \ln 120 = 2,968$ cal/mole to be added to $\mu^o_{\text{vibr.}}$.

[c] The total includes a parcel (3 $RT = 1,860$ cal/mole) corresponding to the heat capacity of the antagonist molecule. Note that the range of variation of this total is small compared to the variation of the energy of affinity ($-\Delta G_i$).

[d] The last column to the right gives the values of $\Delta A = \Delta G_i - (\mu_{\text{vibr.}} + 3\,RT)$. This represents the force pulling the molecules out of the potential well, as indicated in the text and Fig. 7.

creases. Figure 6 gives a schematic view of the situation, showing the increase in affinity ($-\Delta G_i$) as a linear function of the pA$_2$ (according to Eq. 2) and the vibration energy ($\mu_{\text{vibr.}}$) corrected for degrees of freedom (about 120) and thermal capacity (3 RT), as calculated in Table 1. Since free energies are additive, we can propose taking for ΔA the difference between free energy of affinity minus the energy of vibration according to the following equation:

$$\Delta A = \Delta G_i - (\mu_{\text{vibr.}} + 3\,RT) \qquad (5)$$

Since $\mu_{\text{vibr.}}$ is fairly constant, the value of ΔA increases monotonically with the values of pA$_2$, as indicated in the plot of Fig. 7.

We have now to go a step further and relate the data to the time of recovery. From Eq. 1 we can deduce that the inflection point (t_i) of the curves can be calculated from the two constants ΔA and k by the equation

$$t_i = \frac{\ln \Delta A}{k}$$

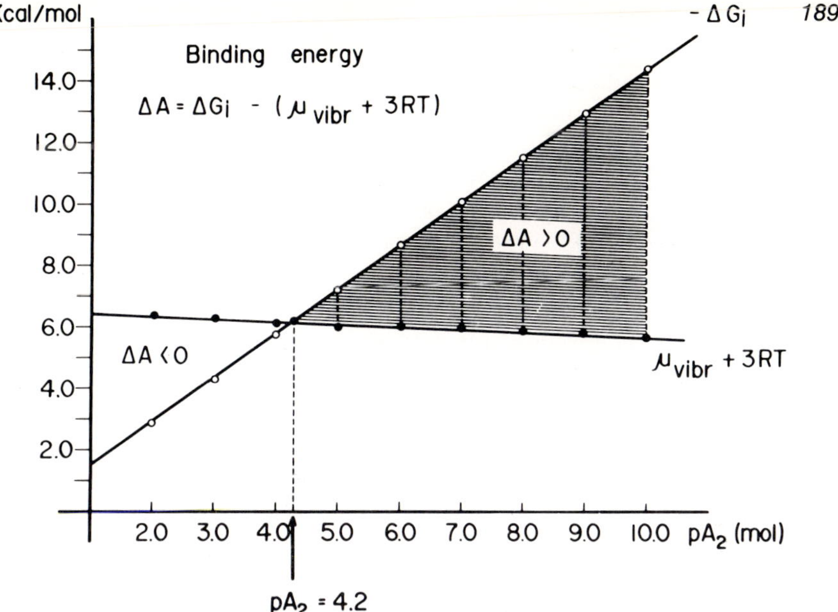

FIG. 6. Binding energy ($\Delta A = \Delta G_i - \mu_{vibr.} - 3\,RT$) of the antagonist molecule in the potential well of Fig. 4. The line indicating the energy of affinity ($-\Delta G_i$) in function of pA$_2$ crosses the line indicating the energy of vibration + heat capacity ($\mu_{vibr.} + 3\,RT$) at a point corresponding to pA$_2$ = 4.2. Below this value $\Delta A < 0$ and above that point $\Delta A > 0$ and increases monotonically with the values of pA$_2$.

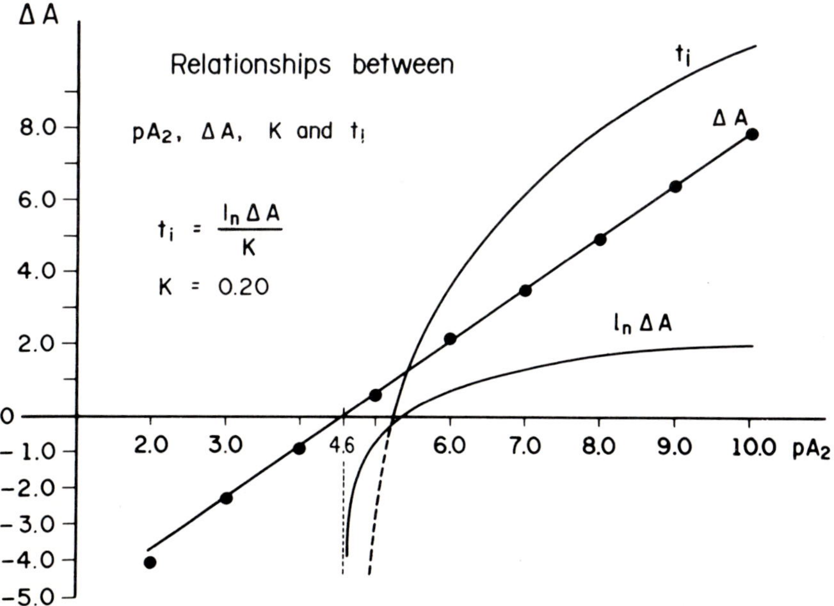

FIG. 7. Relationships between inflexion point (t_i) of equation I, the values of ΔA as indicated in Fig. 6, and the values of pA$_2$ below and above pA$_2$ = 4.6 (w/v) = 4.2 (molar). Note that between pA$_2$ = 7.0 and 10.0, the values of t_i increase linearly with the values of pA$_2$.

The plot of this equation, as related to the values of pA_2 and taking $k = 0.20$, is given in Fig. 7. It can be seen that over a certain range the values of t_i run parallel to the values of pA_2 (from 7.0 to 9.0), which lends support to the previous contention that the time of recovery can constitute a good estimation of the potency of the antihistaminic action (see Rocha e Silva and Beraldo, 1948, and Beraldo and Rocha e Silva, 1949).

An interesting verification of the foregoing deductions was achieved by recalculating a series of parameters of antihistaminic action on guinea pig ileum that have been obtained in our laboratory, starting in 1969, in collaboration with Dr. Abílio Antonio, and which remained unpublished until now. At that time we were investigating the possibility of correlating the values of pA_2 to the values of time of recovery (t_i) for as many drugs as possible, showing weak or strong antihistaminic activity. To calculate such a correlation, the pA_2 was determined indirectly, using the equation

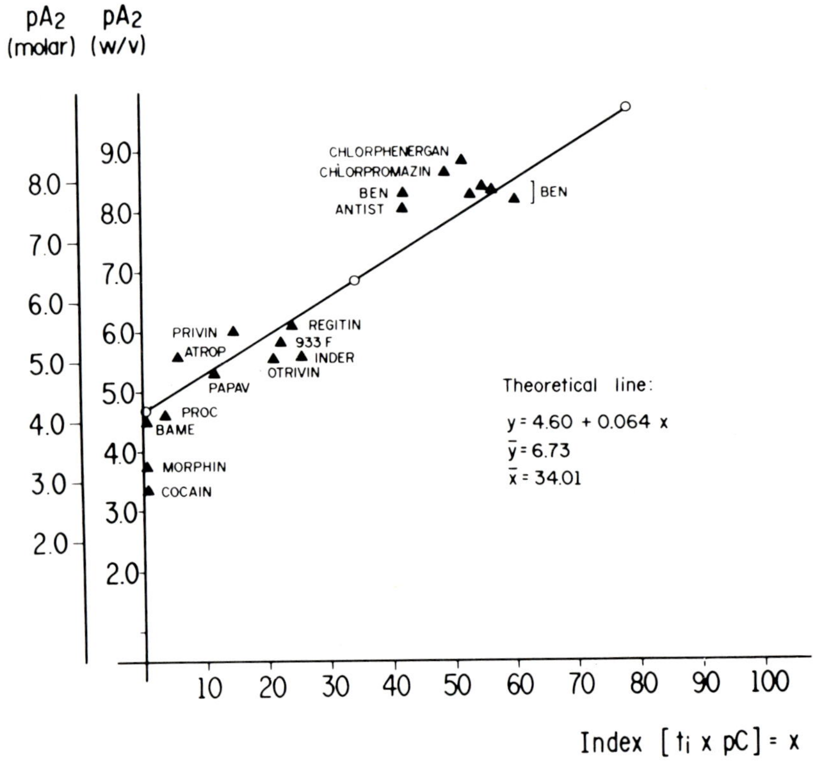

FIG. 8. Correlation between the values of pA_2 of different antagonists toward histamine and time of recovery (t_i) × colog of concentration (pC) applied during 1 min and then washed out. The theoretical line $Y = 4.6 + 0.064 x$ indicates for $x = 0$ the intercept of the line with the coordinate line $pA_2 = 4.6$ (w/v) or $pA_2 = 4.2$ (molar). Below that value, recovery was almost instantaneous after washing out of the antagonist.

$$pA_2 = \log (\beta - 1) - \log I$$

referred to earlier. To obtain a uniform measure of the time of recovery, a large concentration (100 times that corresponding to the pA_2) was applied for a standard interval of time (1 min) and then washed out. The preparation was tested every 2 min with a standard concentration of the agonist (histamine). The time for 37% of recovery (corresponding to t_i) in minutes was multiplied by the reciprocal logarithm of the concentrations applied (pC) and the index, $x = (t_i \times pC)$, was plotted against the values of pA_2 for each substance. The plot in Fig. 8 is a preliminary result obtained with a variety of compounds, displaying weak or strong competitive antagonism toward histamine.

CONCLUDING REMARKS

The plots presented in Figs. 6 and 8 show a remarkable coincidence. Theoretically, the line indicating the energy of vibration ($\mu_{\text{vibr.}} + 3\,RT$) as a function of increasing values of pA_2 ($-\log k_i$) crosses the line of the increase of free energy of affinity ($-\Delta G_i$) at a point corresponding to $pA_2 = 4.2$ (molar), thus indicating that below such a level of pA_2 the energy of binding is fully counterbalanced by the energy of vibration ($\Delta A < 0$). Above that level of pA_2 the values of $\Delta A > 0$ increase monotonically with the values of pA_2. The remarkable coincidence was obtained by calculating the experimental line as represented in Fig. 8, using a number of experimentally obtained data, with compounds showing weak or strong antihistaminic activity. The intercept of the line calculated by the mean-square method gave a limiting value of $pA_2 = 4.2$ (molar), below which no binding (instantaneous recovery) would be expected. Above such a pA_2 there was a linear correlation between pA_2 and the recovery index ($t_i \times$ cologarithm of the concentration applied). The coincidence, of course, was that we got *exactly* the same value for the experimental data and the theoretical value. A simple approximation would be acceptable. Anyway, we can emphasize the experimental value and say that a $pA_2 = 4.2$ calculated as binding energy ($2.3\,RT\,pA_2 = 5.989$ Kcal/mole) would give a rough approximation for the binding energy that keeps the molecule bound to the receptor site.

Experiments in progress in our laboratory, with Dr. Abílio Antonio and Miss Maria Aparecida Funayama, endeavor to verify this relation with other antihistamines and in other biological systems.

ACKNOWLEDGMENTS

This work was supported by grants from the São Paulo State Research Foundation (FAPESP) and the Brazilian National Research Council (CNPg).

REFERENCES

Ariëns, E. J., ed. (1964): *Molecular Pharmacology,* Academic Press, Inc., New York.

Ariëns, E. J., and Simonis, A. M. (1967): Cholinergic and anti-cholinergic drugs: Do they act on common receptors? *Ann. N.Y. Acad. Sci.* 144:842–867.

Baker, B. R. (1962): Non-classical antimetabolites. VII. The bridge principle of specificity with exo-alkylating irreversible inhibitors. *J. Med. Pharm. Chem.* 5:654–656.

Baker, B. R. (1964): Factors in the design of active site-directed irreversible inhibitors. *J. Pharm. Sci.* 53:347–364.

Beraldo, W. T., and Rocha e Silva, M. (1949): Biological assay of antihistaminics, atropine and antispasmodics upon the guinea pig gut. *J. Pharmacol. Exp. Ther.* 97:388–398.

Ritchie, J. M., and Greengard, P. (1966): On the mode of action of local anesthetics. *Annu. Rev. Pharmacol.* 6:405–430.

Rocha e Silva, M. (1969): A thermodynamic approach to problems of drug antagonism. I. The "Charnière Theory." *Eur. J. Pharmacol.* 6:294–302.

Rocha e Silva, M. (1970): A thermodynamic approach to problems of drug antagonism. II. A microphysical model of the phenomenon of recovery. *Physiol. Chem. Phys.* 2:503–515.

Rocha e Silva, M., and Beraldo, W. T. (1948): Dynamics of recovery and measure of drug antagonism. Inhibition of smooth muscle by lysocithin and antihistaminics. *J. Pharmacol. Exp. Ther.* 93:457–469.

Rocha e Silva, M., and Fernandes, F. (1974): A thermodynamic approach to problems of drug antagonism. Measurements of the parameters of affinity (pA_2) of antihistaminics and a β-haloalkylamine under varying experimental conditions. *Eur. J. Pharmacol.* 25:231–240.

Rocha e Silva, M., Fernandes, F., and Antonio, A. (1972): Influence of temperature on recovery of inhibition by antihistaminics and β-haloalkylamines toward histamine. *Eur. J. Pharmacol.* 17:333–340.

Steinberg, I. Z., and Scheraga, H. A. (1963): Entropy changes accompanying association reactions of proteins. *J. Biol. Chem.* 238:172–181.

Stewart, J. M. (1974): Tachyphylaxis to angiotensin. In: *Angiotensin,* edited by I. H. Page and F. M. Bumpus, *Handb. Exptl. Pharmakol.* vol. 37, pp. 170–184. Springer-Verlag New York, Heidelberg.

Concepts of Membranes In Regulation and Excitation,
edited by M. Rocha e Silva and G. Suarez-Kurtz.
Raven Press, New York © 1975.

Neurotransmitter Release by the Toxin of Brazilian Scorpion Venom (*Tityus serrulatus* Lutz e Mello)

A. P. Corrado, C. R. Diniz,* and A. Antonio

*Department of Pharmacology and *Department of Biochemistry, Faculty of Medicine of Ribeirão Preto, University of São Paulo, 14100 Ribeirão Preto, S.P., Brasil*

INTRODUCTION

Direct and indirect evidence discussed in this chapter strongly suggests that the toxin of scorpion venom is able to release both acetylcholine and norepinephrine from depots in the autonomic and somatic nervous systems. The evidence also points to the nerve ending as the site of toxin action.

CHOLINERGIC ACTIONS

Scorpion venom contracts the smooth muscle of guinea pig and rat ileum; the effect is potentiated by neostigmine, blocked by atropine, and partially antagonized by morphine. The ganglion-blocking agent hexamethonium does not interfere with the contraction elicited by the venom (Diniz and Valeri, 1959; Cunha Melo, Freire-Maia, Tafuri, and Maria, 1973).

Scorpion venom also stimulates the neuromuscular junction of the frog (Benoit and Mambrini, 1967; Katz and Edwards, 1972); the venom elicits spontaneous twitches of the rat phrenic nerve-diaphragm preparation and increases the amplitude of the responses to supramaximal pulses; these effects are not seen in curarized and denervated preparations, are calcium dependent, and are abolished by magnesium excess and neomycin (Vital Brazil, Neder, and Corrado, 1973). As shown in Table 1, the addition of an excess of calcium causes reversion of the neuromuscular blockade produced by magnesium and neomycin.

It has been shown that the cholinergic effects of scorpion venom are not the result of anticholinesterase activity, because the venom does not decrease the rate of acetylcholine hydrolysis (Diniz and Gonçalves, 1956; Vital Brazil et al., 1973). Since the bath fluid of preparations submitted to the venom contains a substance that contracts guinea pig ileum and decreases cat blood pressure the cholinergic effects must be due to a release of acetylcholine from the postganglionic and somatic nerve endings. These effects of the bath fluid are better shown in neostigmine-treated preparations

TABLE 1. *Effect of magnesium excess, neomycin, and d-tubocurarine on release of acetylcholine by scorpion venom (10^{-5} g/ml) from isolated rat diaphragm preparation.*[a]

Treatment ($\times 10^{-4}$ g/ml)	CaCl$_2$ excess (3.6×10^{-4} g/ml)	N	Acetylcholine (ng) Mean $\pm$ SEM
Control	−	9	54.2 ± 5.8
MgCl$_2$ (4.1)	−	3	0
MgCl$_2$ (4.1)	+	3	59.2 ± 0.8
Neomycin (4)	−	5	0
Neomycin (4)	+	3	57.5 ± 0.8
d-Tubocurarine (0.03)	−	5	64.5 ± 7.9

[a] No release of acetylcholine was seen in denervated diaphragm or in calcium-free bathing medium.

and are completely blocked by atropine (Diniz and Torres, 1964; 1968; Vital Brazil et al., 1973). Moreover, the amount of the active material released increases with the time of incubation (Diniz and Torres, 1968). The venom also releases acetylcholine from slices of rat brain (Gomez, Dai, and Diniz, 1973), as can be seen in Fig. 1.

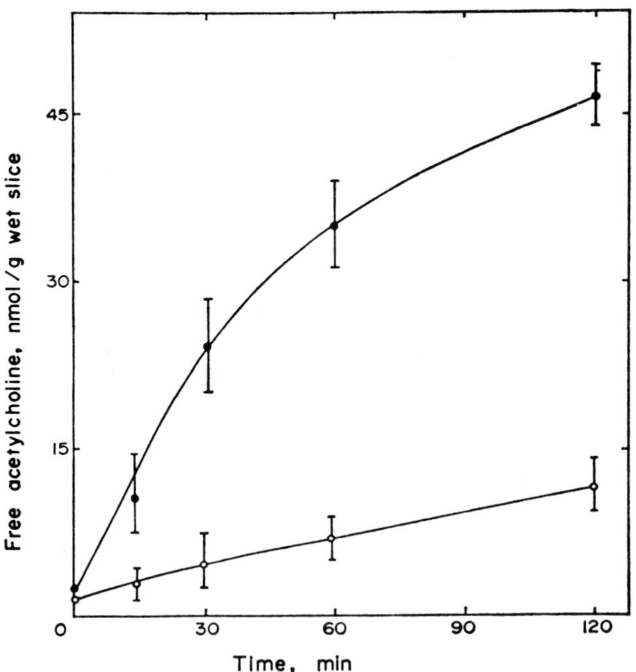

FIG. 1. The influence of incubation time on the release of acetylcholine from slices of rat brain by scorpion venom. The incubation mixture contained, in mM: NaCl, 136; KCl, 2.7; CaCl$_2$, 1.35; NaH$_2$PO$_4$, 0.36; NaHCO$_3$, 12; glucose, 5.5; physostigmine, 0.01; pH = 7.4. ○ without venom; ● with 40 μg (2.0 μM) venom. Each point represents the mean ± SD of 3 experiments.

The acetylcholine-releasing effect of scorpion venom is completely blocked by tetrodotoxin (Gomez et al., 1973). It has been well demonstrated that scorpion venom produces a sustained depolarization of the nerve membrane (Narahashi, Shapiro, Deguchi, Scuka, and Wang, 1972; Schmitt and Schmidt, 1972), which might explain the long duration of pharmacological action.

ADRENERGIC ACTIONS

When added to rabbit or rat duodenum immersed in atropine containing Tyrode solution, scorpion venom produces relaxation (Ramos and Corrado, 1954; Freire-Maia and Ferreira, 1961; Cunha Melo et al., 1973).

More consistent adrenergic-like effects of the venom are seen in the isolated guinea pig heart, using Langendorff's technique. The typical effect of the toxin is a short-lasting bradycardia followed by a conspicuous increase in the force and frequency of the heart beats; phosphorylase *a* activity is also increased; as might be expected, the bradycardia is blocked by atropine and potentiated by neostigmine (Corrado, Antonio, and Diniz, 1968). The cardiac stimulation and enzyme activation are completely blocked by either propranolol or bretylium and are absent in the hearts of reserpine-treated guinea pigs (Table 2).

Although the effects of scorpion venom are quite similar to those evoked by nicotine, they are not affected by hexamethonium (Fig. 2). The differing influences of hexamethonium on the bradycardic action of nicotine and scorpion venom can also be shown in reserpine-treated preparation (Fig. 3).

All the aforementioned effects of scorpion venom observed upon the isolated heart are easily reproduced in either the spontaneously beating or

TABLE 2. *Effect of scorpion venom (20 µg) on isotonic force of contraction and phosphorylase a activity of perfused guinea pig heart*

		Positive inotropic effect			
Treatment (µg/min)	N	Maximal effect (% of control) Mean ± SEM	Total duration (min) Mean ± SEM	Duration of bradycardia[a] (sec) Mean ± SEM	Phosphorylase a activity (%) Mean ± SEM
Control	5	282 ± 37.5	16.4 ± 1.9	72 ± 4.8	68.9 ± 5.6
Bretylium (100)	4	—	—	116 ± 5.7	32.6 ± 6.4
Propranolol (0.5)	4	—	—	128 ± 5.4	31.8 ± 4.7
Hexamethonium (50)	5	196 ± 27.6	14.7 ± 2.3	67 ± 4.9	70.1 ± 5.9
Reserpine[b]	6	—	—	186 ± 6.3	34.2 ± 2.9

[a] Bradycardia always precedes the stimulating effect; it is potentiated by neostigmine and blocked by atropine.

[b] Guinea pigs treated with reserpine received a total of 2 mg/kg equally divided in 2 doses given i.p. 48 and 24 hr before the experiments. The dash (−) denotes absence of the stimulating effect of scorpion venom.

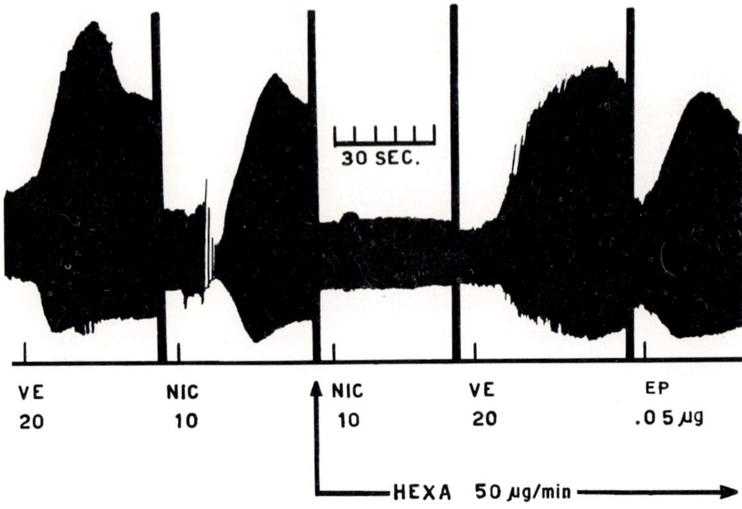

FIG. 2. Isotonic contractions of the isolated heart of the normal guinea pig. The numbers represent total doses in micrograms. The infusion of hexamethonium (HEXA), started at the arrow, blocks both effects of nicotine (NIC) but not the cardiac stimulation by the venom (VE) or epinephrine (EP). After the effect of each drug, the kymograph was stopped for 20 min at the vertical bars.

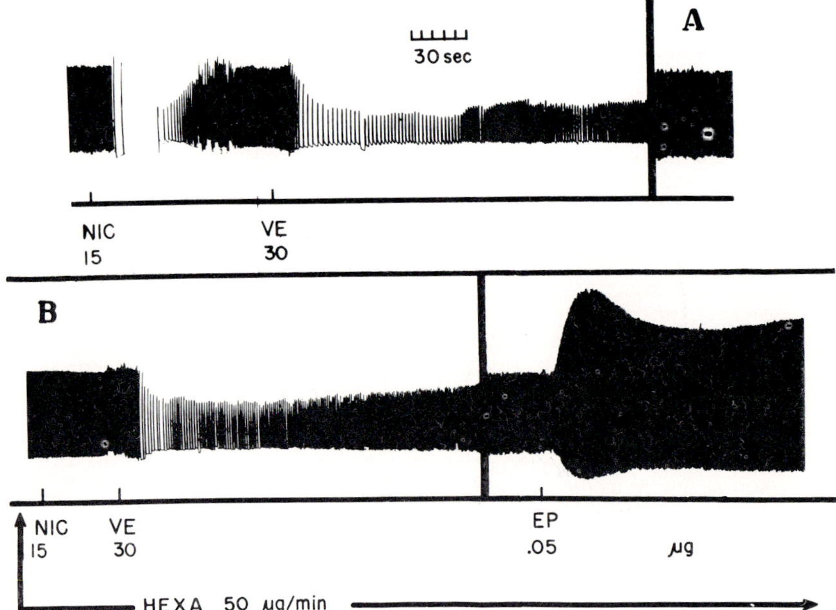

FIG. 3. Isotonic contractions of the isolated heart of the reserpine-treated guinea pig. Drugs and dosages as in Fig. 2; treatment with reserpine as in Table 2. Observe that bradycardia is the only effect seen with both nicotine and venom (*upper panel*, A) and that hexamethonium (*lower panel*, B) does not block the response to venom.

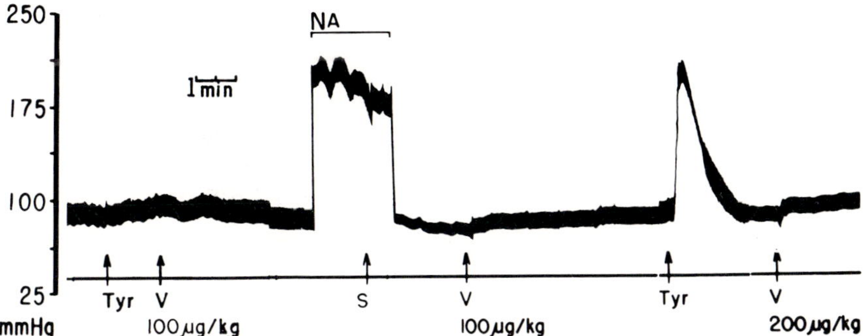

FIG. 4. Blood pressure record of a reserpinized female cat (3.0 kg) treated with atropine sulfate (1 mg/kg). The hypertensive response to 1 mg of tyramine (Tyr) but not that of venom (V) is restored following the infusion of 0.5 mg/kg of noradrenaline (NA). At S the kymograph was stopped for 10 min. Reserpine was administered in 3 i.p. injections of 0.5 mg/kg, 72, 48, and 24 hr before the experiment.

electrically driven isolated atria of the guinea pig, and are consistent with an indirect release of norepinephrine from the postganglionic nerve terminals (Almeida and Antonio, *unpublished*).

Additional support for our conclusions have been obtained by *in vivo* experiments. Thus the intravenous injection of scorpion venom in anesthetized dogs, cats, and rabbits produces a prolonged increase in blood pressure, which is usually preceded by a slight and short-lived hypotension; a decrease in the heart rate is also seen during the hypertensive effect. Atropine potentiates the hypertension and prevents the initial fall in blood pressure as well as the bradycardia produced by the venom. The pressure changes elicited by the venom are not affected by doses of hexamethonium that completely block the responses to nicotine. The hypertensive response is blocked by alpha-sympatholytic agents and is absent in reserpine-treated or guanethidine-treated animals (Corrado, Riccioppo Neto, and Antonio, 1974). Cocaine prevents the effect of tyramine in doses that potentiate the hypertensive response to the venom. Moreover, the infusion of norepinephrine in reserpine-treated animals restores the effect of tyramine but not the effect of venom (Fig. 4).

Following the administration of the venom there is also an augmentation of the hypertensive response to the bilateral carotid occlusion and, as shown in Fig. 5, to injected norepinephrine and tyramine (Corrado et al., 1974).

More direct evidence has been obtained recently by Moss, Colburn, and Kopin (1974*a*) and Moss, Thoa, and Kopin (1974*b*). They showed a release of ^{3}H-norepinephrine by the toxin of the scorpion *Leiurus quinquestriatus*. Preliminary experiments (Langer, Adler-Graschinsky, Almeida, and Diniz, *unpublished*) have shown that in addition to the release of ^{3}H-norepinephrine, the venom potentiates the release of neurotransmitter by the nervous impulse.

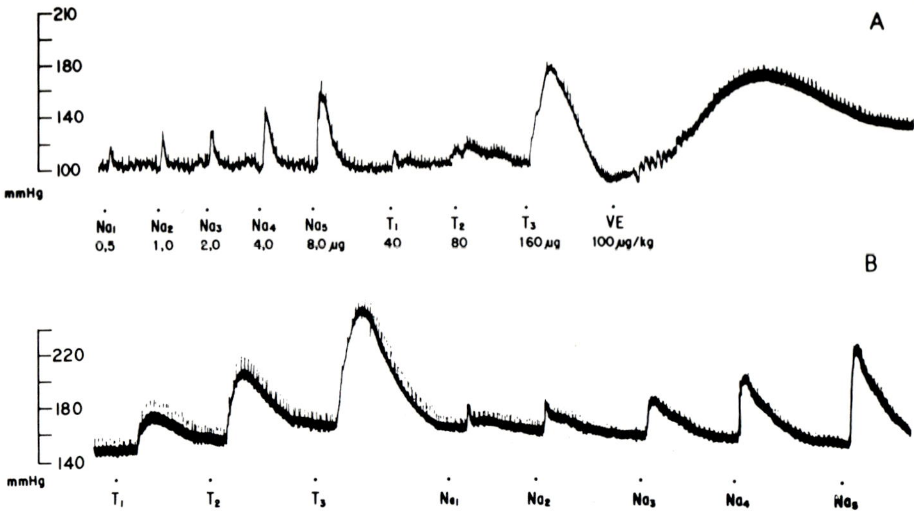

FIG. 5. Blood pressure record of a female dog (9.5 kg) treated with atropine sulfate (1 mg/kg). The upper panel (A) shows control responses to increasing doses of nor-adrenaline (NA) and tyramine (T); the venom was then injected and 10 min later (*lower panel,* B) the injections of tyramine and noradrenaline were repeated. It can be seen that the action of both amines was potentiated by the venom.

CONCLUSIONS

The results discussed here clearly show that scorpion venom releases both acetylcholine and norepinephrine by acting at the postganglionic nerve terminals. Similar release of acetylcholine is observed at the neuromuscular junction and in incubated slices of rat brain. The venom also releases catecholamines from the adrenal gland (Henriques, Gazzinelli, Diniz, and Gomez, 1968).

The mechanism of release is quite similar to that induced by electrical stimulation of the nervous impulse. Accordingly, the venom produces long-lasting action potentials (Narahashi et al., 1972; Schmitt and Schmidt, 1972), which might explain the long duration of its pharmacological effects. The releasing effect is partially dependent on the presence of calcium in the bath fluid and is inhibited by tetrodotoxin (Vital Brazil et al., 1973; Gomez et al., 1973; Moss et al., 1974a). Finally, it should be mentioned that scorpion toxin increases the synthesis of acetylcholine (Gomez et al., 1973) and norepinephrine (Moss et al., 1974b), potentiates the pharmacological actions of exogenous and endogenous norepinephrine, and tyramine (Corrado et al., 1974), and enhances the release of transmitters via nerve stimulation (Langer et al., *unpublished*).

ACKNOWLEDGMENT

This work was supported in part by research grants from FAPESP, São Paulo, Brazil.

REFERENCES

Benoit, P. R., and Mambrini, J. (1967): Action of scorpion venom on the neuromuscular junction of frogs. *J. Physiol.* (Paris) 59:348.

Corrado, A. P., Antonio, A., and Diniz, C. R. (1968): Brazilian scorpion venom (*Tityus serrulatus*), an unusual sympathetic postganglionic stimulant. *J. Pharmacol. Exp. Therap.* 164: 253–258.

Corrado, A. P., Riccioppo Neto, F., and Antonio, A. (1974): The mechanism of the hypertensive effect of Brazilian scorpion venom (*Tityus serrulatus* LUTZ E MELLO). *Toxicon* 12:145–150.

Cunha Melo, J. R., Freire-Maia, L., Tafuri, W. L., and Maria, T. A. (1973): Mechanism of action of purified scorpion toxin on the isolated rat intestine. *Toxicon* 11:81–84.

Diniz, C. R., and Gonçalves, J. M. (1956): Some chemical and pharmacological properties of Brazilian scorpion venoms. In: *Venoms,* edited by E. E. Buckley and N. Porges, p. 131. AAAS, Washington, D.C.

Diniz, C. R., and Torres, J. M. (1964): Release of acetylcholine by scorpion venom. *Ciencia Cult.* 16:197.

Diniz, C. R., and Torres, J. M. (1968): Release of an acetylcholine-like substance from guinea pig ileum by scorpion venom. *Toxicon* 5:277–281.

Diniz, C. R., and Valeri, V. (1959): Effects of a toxin present in a purified extract of telsons from the scorpion *Tityus serrulatus* on smooth muscle preparations and in mice. *Arch. Intern. Pharmacodyn.* 71:1–13.

Freire-Maia, L., and Ferreira, M. C. (1961): Estudo do mecanismo da hiperglicemia e da hipertensão arterial produzidas pelo veneno de escorpião no cão. *Mem. Inst. Oswaldo Cruz* 59:11–12.

Gomez, M. V., Dai, M. E. M., and Diniz, C. R. (1973): Effect of scorpion venom, Tityustoxin, on the release of acetylcholine from incubated slices of rat brain. *J. Neurochem.* 20:1051–1061.

Henriques, M. C., Gazzinelli, G., Diniz, C. R., and Gomez, M. V. (1968): Effect of the venom of the scorpion *Tityus serrulatus* on adrenal gland catecholamines. *Toxicon* 5:175–179.

Katz, N. L., and Edwards, C. (1972): The effect of scorpion venom on the neuromuscular junction of the frog. *Toxicon* 10:133–137.

Moss, J., Colburn, R. W., and Kopin, I. J. (1974a): Scorpion toxin induced catecholamine release from synaptosomes. *J. Neurochem.* 22:217–221.

Moss, J., Thoa, N. B., and Kopin, I. J. (1974b): On the mechanism of scorpion toxin-induced release of norepinephrine from peripheral adrenergic neurons. *J. Pharmacol. Exp. Ther.* 190:39–48.

Narahashi, T., Shapiro, B. I., Deguchi, T., Scuka, M., and Wang, C. M. (1972): Effects of scorpion venom on squid axon membranes. *Am. J. Physiol.* 222:850–857.

Ramos, A. O., and Corrado, A. P. (1954): Efeito hiperpiético do veneno de escorpião *T. serrulatus* e *T. bahiensis. Anais Fac. Med. USP* 28:81–98.

Schmitt, O., and Schmidt, H. (1972): Influence of calcium ions on the ionic currents of nodes of Ranvier treated with scorpion venom. *Arch. Ges. Physiol.* 333:51–61.

Vital Brazil, O., Neder, A. C., and Corrado, A. P. (1973): Effects and mechanism of action of *Tityus serrulatus* venom on skeletal muscle. *Pharmacol. Res. Comm.* 5:137–150.

Concepts of Membranes in Regulation and Excitation,
edited by M. Rocha e Silva and G. Suarez-Kurtz.
Raven Press, New York © 1975.

Competitive Antagonism between Calcium and Aminoglycoside Antibiotics in Skeletal and Smooth Muscles

A. P. Corrado, W. A. Prado, and I. Pimenta de Morais

Department of Pharmacology, Faculty of Medicine of Ribeirão Preto, University of São Paulo, 14100 Ribeirão Preto, S.P., Brazil

INTRODUCTION

The aminoglycoside antibiotics are basic substances elaborated by actinomyces of the genus *Streptomyces* and *Micromonospora*. Among them streptomycin, dihydrostreptomycin, neomycin, paromomycin, aminosidin, and gentamycin are of major clinical importance because of their wide therapeutic use. Their structures are similar (Fig. 1) and they induce qualitatively similar toxic effects which—depending on dosage, method of administration, and duration of treatment—can lead to an acute or chronic toxic picture.

When administered to laboratory animals in therapeutic doses for long periods of time, aminoglycoside antibiotics may cause selective lesions in the labyrinth of the ear, both in the cochlear and vestibular portions, thus inducing, respectively, progressive and irreversible deafness and/or equilibrium and coordination alterations and nystagmic disturbances.

Acute toxicity can occasionally be observed after intravenous or intraperitoneal administration of doses large enough to produce a high concentration in the bloodstream. Although the intravenous route is of restricted therapeutic use and its importance is confined to the study of experimental acute toxicity in laboratory animals, several cases of acute toxicity in man have been reported after intraperitoneal instillation of these antibiotics (Pridgen, 1956; Case Report No. 190, 1957; Webber, 1957; Ferrara and Phillips, 1957; Engel and Denson, 1957; Middleton, Morgan, and Moyers, 1957; McCorkle, 1958; Pittinger and Long, 1958; Jones, 1959; Kownacki and Serlin, 1960; Richet, Ducrot, and Delzant, 1960; Mullet and Keats, 1961; Fisk, 1961; Benz, Lunn, and Foldes, 1961; Bush, 1961; Foldes, 1963; Pandey, 1964). Such toxicity is characterized in both laboratory animals and man by sudden cardiorespiratory failure (Corrado, 1963; Pittinger and Adamson, 1972).

The first experimental studies describing in detail the symptoms of acute

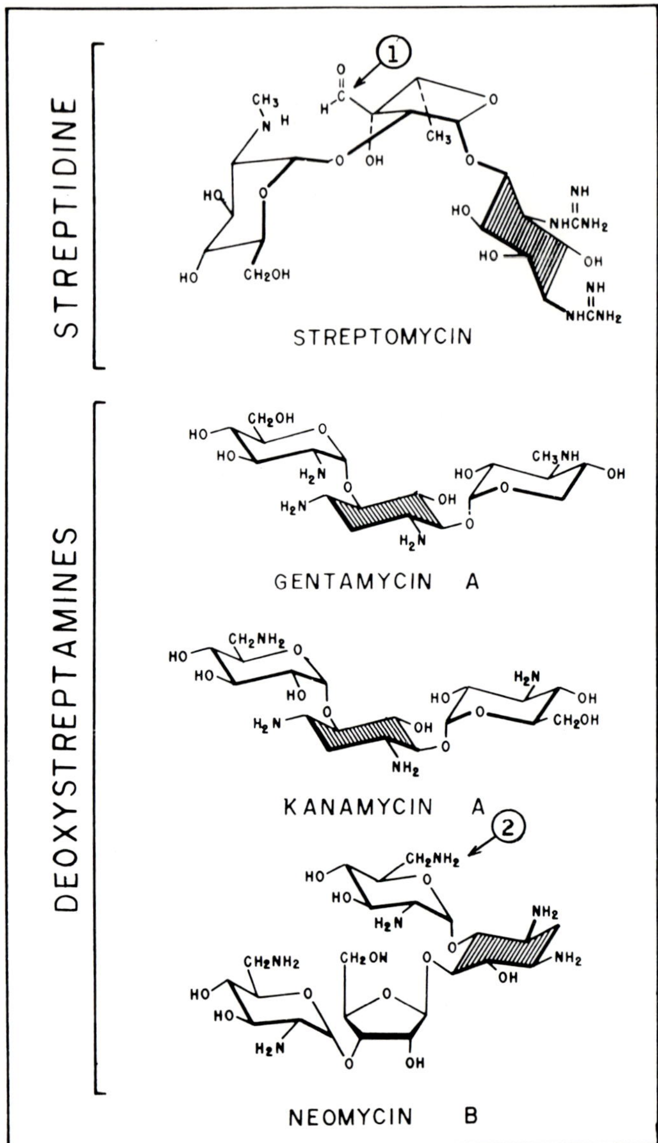

FIG. 1. Chemical structures of aminoglycoside antibiotics. The common cyclitolic moiety (*shaded areas*) is an inositol in which the -OH groups are replaced in positions 1 and 3 by guanidine (streptidine group) or by -NH$_2$ (2-deoxistreptamine group). Catalytic dehydrogenation at (1) furnishes dihydrostreptomycin; substitution of -NH$_2$ for -OH in position (2) gives paromomycin (aminosidin).

toxicity caused by streptomycin and dihydrostreptomycin attributed the hypotension and respiratory failure to direct depression of the vasomotor and respiratory centers (Molitor and Graessle, 1950). Subsequent studies, however, have revealed that such effects are the result of blocking actions on the ganglionic and neuromuscular synapses (Corrado, 1963). More recently it has been demonstrated that other antibiotics of the group also cause neuromuscular actions and cardiovascular effects with characteristics similar to those of streptomycin and dihydrostreptomycin, probably by the same mechanism of action (Pittinger and Adamson, 1972).

Vital Brazil and Corrado (1957) and Corrado and co-workers (Corrado and Ramos, 1958, 1960; Corrado, Ramos, and Escobar, 1959) were the first to show the strong similarity between the neuromuscular blockages caused by the aminoglycoside antibiotics and by Mg^{++}, and to demonstrate that the antidote drugs capable of blocking the toxic effects are neostigmine and Ca^{++}—the effects of neostigmine being weak and irregular and those of Ca^{++} being more efficient. Later studies confirmed that calcium is the antagonist of choice, even in cases of acute toxicity caused by the aminoglycoside antibiotics in man (Jones, 1959; Mullet and Keats, 1961).

COMPETITIVE ANTAGONISM AT THE NEUROMUSCULAR JUNCTION

The important role played by calcium in the process of neuromuscular transmission is well established. It has been shown that the liberation of acetylcholine depends on adequate extracellular calcium concentration (del Castillo and Stark, 1952). The significance of the Ca^{++} level in the extracellular medium was demonstrated by the study of calcium-magnesium antagonism at the neuromuscular junctions (del Castillo and Engbaek, 1954). On the assumption that the quantity of mediator liberated by the nervous impulse would depend on the relative concentrations of calcium and magnesium in the extracellular medium, the suggestion was made that there is a competitive interaction between those ions at the presynaptic level of the neuromuscular junction (del Castillo and Engbaek, 1954). This hypothesis was later confirmed by Jenkinson (1957) and Dodge and Rahamimoff (1967) through the mathematical analysis of the effects induced by Ca^{++} and Mg^{++} on the quantitative liberation of acetylcholine in the end plates of frog skeletal muscles. It was postulated that these ions interact with specific receptor sites—designated X—of the presynaptic membrane to reversibly form CaX complexes which somehow aid in the process of liberating the transmitter (Fatt and Katz, 1952; del Castillo and Katz, 1954a,b). MgX complexes are also formed reversibly, but with a higher dissociation constant (Dodge and Rahamimoff, 1967); they are ineffective in the release of mediator, a fact that might explain the competitive antagonism observed for these ions.

A similar antagonism between aminoglycoside antibiotics and calcium was also suspected because of the similarity between the myographic patterns described for neuromuscular blockages by Mg^{++} and by the antibiotics. This suggestion was initially strengthened by Elmqvist and Josefsson (1962), who recorded a marked reduction, induced by neomycin, in the frequency of discharge of miniature end-plate potentials evoked by high external concentration of K^+ in isolated frog sartorius muscle. In our laboratory (Prado and Corrado, *to be published*), this suggestion is corroborated by quantitative analysis of the myographic effects induced by streptomycin and kanamycin in isolated rat diaphragm and by the electrophysiological alterations caused by these antibiotics in the end-plate potentials of isolated toad sartorious muscle.

Different doses of antibiotics were added at random to isolated rat phrenic nerve-diaphragm preparations assembled according to the method of Bülbring (1946); the preparations, containing different concentrations of calcium chloride (1.6, 1.8, and 2.0 mM), were immersed in Tyrode-glycose[1] at 37°C. The assay was repeated in 6 preparations for each antibiotic used, and the average values of the neuromuscular blockages were plotted as a function of the logarithm of antibiotic concentrations applied to the bath. Log dose-effect curves were obtained with no statistically significant deviation from parallelism, as shown for kanamycin assays in Fig. 2—a fact that suggests the competitive character of the phenomenon studied (Arunlakshana and Schild 1959). Similar results were obtained with streptomycin.

With the use of calcium-chelating agents such as sodium salicylate and tetracycline we obtained irreversible blockage. With Versene we obtained log dose-effect curves with a significant deviation from parallelism (Fig. 2). Therefore we concluded that a simple complex formation with Ca^{++} does not explain the alterations induced by the antibiotics in the liberation of the mediator, and that a mechanism of competition at the presynaptic neuromuscular level is probably the most plausible explanation.

Additional support for this hypothesis was obtained by application of the mathematical model proposed by Dodge and Rahamimoff (1967) concerning the quantitative relation between calcium and acetylcholine output at the neuromuscular junctions of isolated toad sartorius muscle. According to the technique of Fatt and Katz (1951), this preparation was immersed in a nutrient liquid[2] at room temperature in presence of *d*-tubocurarine (3.0×10^{-6} g/l) and at various concentrations of calcium (1.0, 1.4, 1.8, and 2.2 mM). At each of those concentrations the muscle was submitted to a series of 16 stimuli of 0.2 msec duration and at a frequency of 6/min in the absence or presence of increasing doses of antibiotics. An intracellular microelectrode

---

[1] Tyrode-glycose solution: NaCl, 137.0; KCl, 3.0; $MgCl_2.2\ H_2O$, 1.0; $NaHCO_3$, 12.0; $NaHPO_4$, 0.4; Glycose, 11.0. Concentrations in millimoles per liter.

[2] Nutrient solution: Na^+, 116.55; K^+, 2.50; Cl^-, 117.10; H_2PO_4, 0.45, HPO_4, 2.55. Concentrations in millimoles per liter.

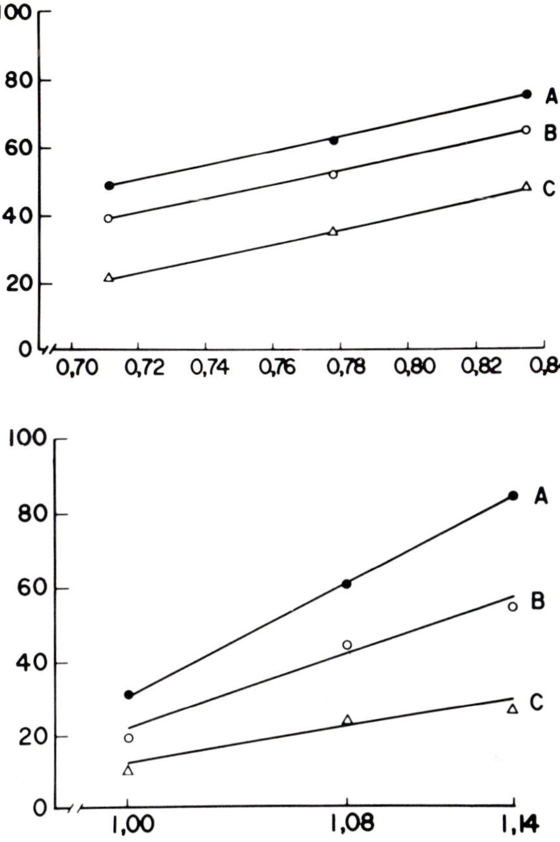

FIG. 2. Log dose-response curves of the percent of neuromuscular block induced by kana-mycin (*upper graph*) and disodium-EDTA (*bottom graph*) as a function of the logarithm of the mM concentrations of these agents. Each point represents the average of 6 experiments performed in the presence of different concentrations of calcium in the bath fluid: 1.6, 1.8, and 2.0 mM at A, B, and C, respectively. (From Prado and Corrado, *unpublished results.*)

was used for recording the end-plate potentials (epp), which were automatically averaged by a Nicolet NC 1072 computer. The double reciprocal plot of the fourth root of the amplitude of the end-plate potentials as a function of the logarithm of external calcium concentration gave straight lines with approximately the same intercept at the ordinates, as can be seen in Fig. 3. As in the case of Mg^{++}, the antibiotics change only the slope without changing the intercept significantly. This graphical representation confirming the competitive antagonism between the aminoglycoside and calcium ions, was made by applying the Lineweaver-Burk double reciprocal transformation of the equation of Dodge and Rahamimoff (1967) for com-

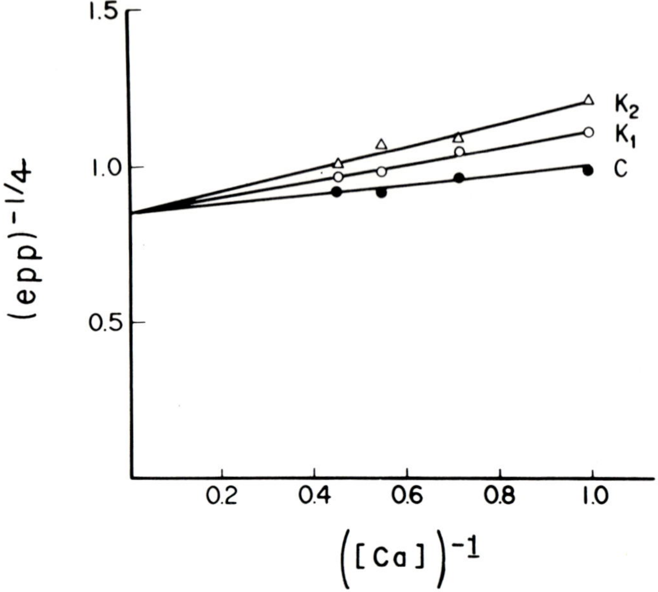

FIG. 3. Double reciprocal plot of the fourth root of the amplitude of the end-plate potentials (epp) as a function of the logarithm of the external calcium concentration (mM) obtained in absence (C) and presence of the two concentrations of kanamycin in the bath fluid: 0.04 and 0.08 mM at K_1 and K_2, respectively. (From Prado and Corrado, *unpublished results.*)

petitive antagonism between calcium and magnesium ions at the neuro-muscular junctions:

$$\text{epp }(m) = k\,(\text{CaX})^4 = k\left[\frac{W\,[\text{Ca}]}{1 + \dfrac{[\text{Ca}]}{K_1} + \dfrac{[\text{Mg}]}{K_2}}\right]^4$$

The amplitude of end-plate potentials (epp) or the quantal content (m) is directly proportional to $(\text{CaX})^4$ complexes; that is, at least four adjacent calcium-binding sites at the terminal have to be occupied simultaneously to give a release of 1 quanta of transmitter; k is the maximal number of quanta the release mechanism is able to liberate at "infinite" calcium concentrations; W is a constant and K_1 and K_2 are the dissociation constants for the CaX and MgX complexes, respectively.

In our experimental conditions, we substitute [Mg] for antibiotics [AB], K_3 being the dissociation constant of the hypothetical interaction between these substances and the X sites of calcium binding at the nerve terminal:

$$\text{epp } (m) = k \ (CaX)^4 = k \left[\frac{W \ [Ca]}{1 + \dfrac{[Ca]}{K_1} + \dfrac{[AB]}{K_3}} \right]^4$$

Taking into account the results obtained with rat diaphragm and toad sartorius muscle, we concluded that aminoglycosides do not change the Ca^{++}-induced maximal response of either preparation, but apparently there is a decrease of the affinity of the calcium ions for their receptor sites at the prejunctional ending, which can be reversed by increasing the extracellular ion concentration. These observations confirm the similarity of the neuromuscular blocks produced by the antibiotics and magnesium ions. They also indicate an interaction of the antibiotics with selective X calcium receptors of the presynaptic membrane, leading to the formation of the ABX complexes that are ineffective for neuromuscular transmission; i.e., there is no more formation of CaX complexes that are reversible and responsible for the neurotransmission.

It should be recalled that streptidine (Fig. 1)—the inositol moiety of streptomycin, considered to be responsible for the toxic effects of the antibiotic (Vital Brazil, Corrado, and Berti, 1957)—showed the same competitive antagonism toward calcium ions, suggesting an effective participation of the cyclitolic portion of the molecule in the neuromuscular blocking activity of the antibiotics.

COMPETITIVE ANTAGONISM IN SMOOTH MUSCLE

Earlier reports refer to the relaxing effect of aminoglycoside antibiotics on intestinal smooth muscle and to the blocking action of these compounds on peristaltic reflexes *in vitro* as well as *in vivo*. Like the action observed in denervated skeletal muscle (Vital Brazil, 1961), the relaxing effect of the antibiotics, as well as that of magnesium is not antagonized by calcium ions; indeed, calcium potentiates the relaxation caused by these agents through its well-known membrane-stabilizing effect.

Recently the effects of aminoglycosides and magnesium ions on calcium-evoked contractions and ^{45}Ca uptake and efflux have been studied (Pimenta de Morais, Corrado, and Suarez-Kurtz; *to be published*) in isolated, depolarized guinea pig ileum preparations, in which contractility is a hyperbolic function of the external calcium ion concentration (Hurwitz and Suria, 1971).

After control dose-response curves were obtained by increasing calcium concentration in the external medium,[3] the guinea pig ileum was treated

---

[3] Calcium-free potassium-Tyrode solution: NaCl, 42; KCl, 98; $NaHCO_3$, 1.2; NaH_2PO_4, 0.04; $MgCl_2$, 1.0; glycose, 5.5; Tris, 5.0; demineralized water, 1 liter. Concentrations in millimoles per liter.

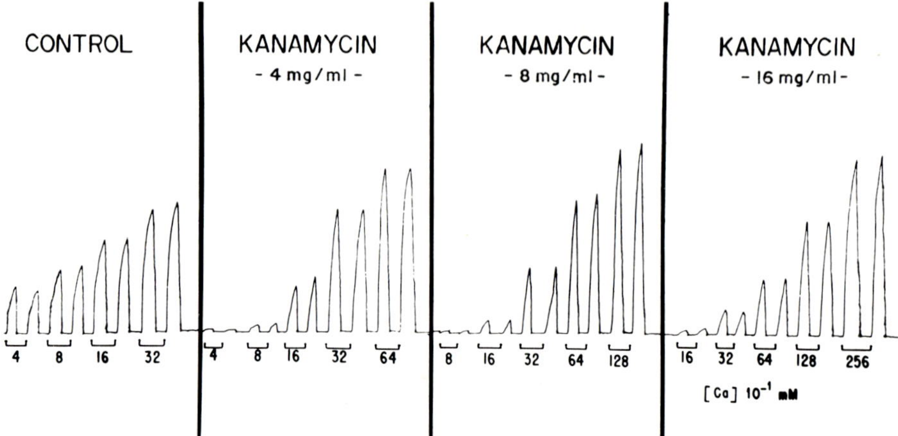

FIG. 4. Isotonic contractions of the isolated guinea pig ileum elicited by increasing external calcium concentration (from 4 to 256 × 10⁻¹ mM) in a calcium-free potassium-Tyrode solution. Note that the inhibitory action of kanamycin is proportional to the concentration of the antibiotic in the bath fluid and can be easily reversed by increasing the external calcium concentration. (From Corrado et al., 1971.)

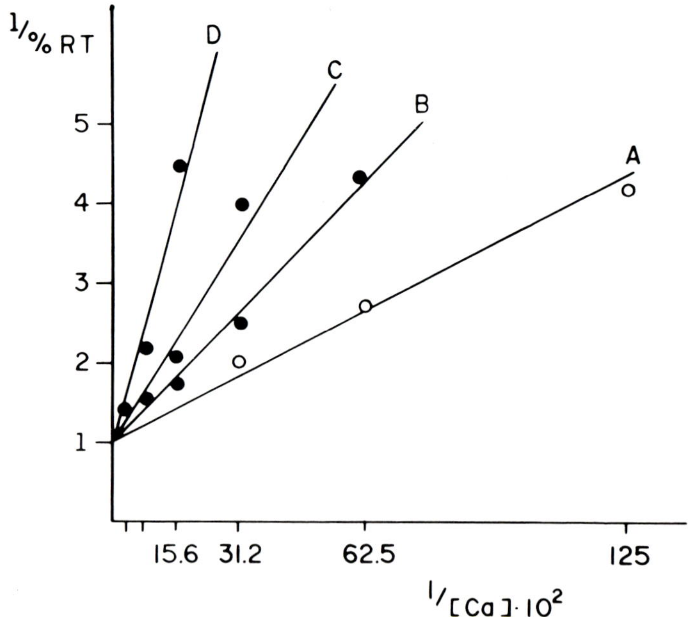

FIG. 5. Double-reciprocal plot of the contractile responses (percent of maximal effect) versus calcium concentration in the extracellular fluid (mM) in absence (A) or presence (B, C, and D) of increasing doses of kanamycin: 4.0, 8.0, and 16.0 mg/ml, respectively. (From Corrado et al., 1971.)

with several concentrations of antibiotics and magnesium ions. In the presence of these antagonistic agents, a fourfold increase in calcium concentration was required to obtain equivalent contractile responses (Fig. 4). This inhibitory effect was reversible, the responsiveness of the preparation being restored to control levels 15 min after removal of the antibiotics or the magnesium from the medium. The straight lines—obtained by the method of the double-reciprocal plot (Rocha e Silva, 1959) for calcium concentration against the contractile effects in the presence and absence of the inhibitory agents—have a common intercept at the ordinate, suggesting a competitive antagonism (Fig. 5). The pA_2 index—defined by Schild (1947) as a measure of the relative potencies of competitive agents—shows that on a molar basis the inhibitory potency decreases in the following way ($pA_2 \pm$ SE; number of experiments): gentamycin (3.95 ± 0.05; $n = 6$) > magnesium (3.73 ± 0.08; $n = 6$) > neomycin (3.40 ± 0.07; $n = 6$) > streptomycin (3.19 ± 0.04; $n = 21$) = streptidine (3.14 ± 0.06; $n = 8$) > kanamycin (2.20 ± 0.10; $n = 4$).

In contrast, calcium-chelating agents such as Versene, sodium salicylate, and tetracycline (Albert, 1965) did not display the characteristics of com-

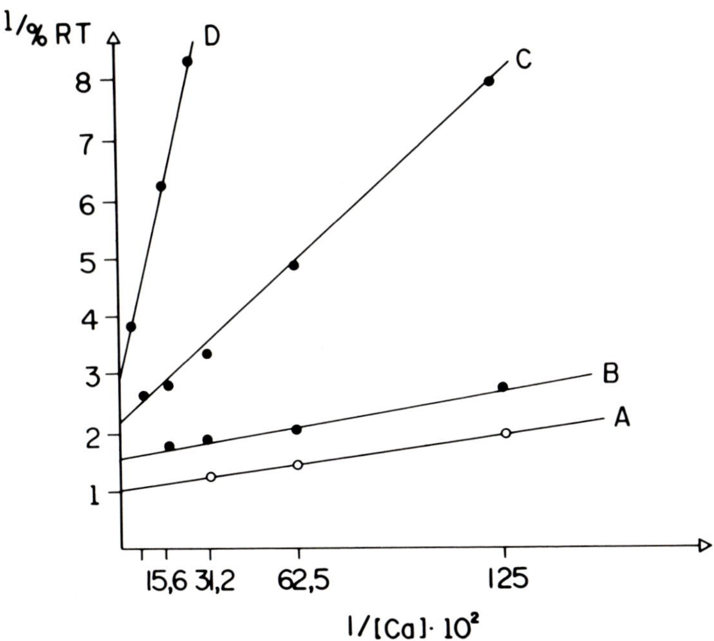

FIG. 6. Double-reciprocal plot of the contractile responses (% of maximal effect) versus calcium concentration in the extracellular fluid (mM) in absence (A) or presence (B, C, and D) of increasing doses of tetracycline: 0.4, 0.8, and 1.6 mg/ml, respectively. (From Corrado et al., 1971.)

petitive antagonism, since both the slope and the intercept at the ordinates of the straight lines relating the reciprocals of calcium concentrations to the reciprocals of the evoked contractions were altered (Fig. 6).

These data confirm the existence of a calcium-saturable system in the smooth muscle cells of the intestine, as suggested by Hurwitz and Joiner (1969) for the longitudinal muscle of the guinea pig ileum. According to the three-step model proposed by Hurwitz and Suria (1971) for activation of contractility in smooth muscle, the blockade of calcium-elicited contractions could be attributed to a competition between the antibiotics and calcium ions for the binding sites on the transport system that carries the calcium ions across the cell membrane. This would result in decreased calcium influx, reduction in the level of the myoplasmic calcium concentration, and consequently less activation of the contractile machinery.

The effects of one of the aminoglycosides — streptomycin — on ^{45}Ca binding and efflux in the same biological preparation (Pimenta de Morais, Corrado, and Suarez-Kurtz; *to be published*), indirectly support this interpretation. In fact, streptomycin elicited a pronounced dose-dependent fall in the uptake of ^{45}Ca, with reduction of both the fast and slow components of cellular ^{45}Ca spaces. Data from nine paired preparations indicating this effect are shown in Table 1. Streptomycin also affected the rate of ^{45}Ca efflux; addition of the antibiotic to the wash-out solutions markedly increased the slope of the slow component of the desaturation curve. When these data were presented as changes in the rate coefficient of ^{45}Ca efflux, it was observed that the effects of streptomycin were maximal in the first sample

TABLE 1. *Effect of streptomycin on ^{45}Ca uptake in depolarized guinea pig ileum[a]*

Experiment	^{45}Ca space control	Streptomycin (ml/g)
		(1.15 mM)
1	3.05	2.48
2	5.39	4.14
3	2.88	2.10
4	1.64	1.25
		(2.30 mM)
1	2.26	1.45
2	2.26	1.54
		(4.60 mM)
1	2.69	1.40
2	3.92	2.70
3	3.09	1.63

[a] From Pimenta de Morais, Corrado, and Suarez-Kurtz, *unpublished results.*

collected, declining thereafter, although the rate of efflux was kept above control values throughout the exposure to the antibiotic.

The increase of ^{45}Ca efflux by aminoglycosides more likely reflects displacement of bound ^{45}Ca than either an increase in membrane permeability to calcium ions or activation of the postulated energy-dependent mechanism for calcium extrusion in intestinal smooth muscle (Hurwitz and Suria, 1971). Goodman, Weiss, and Adams (1974) also observed an increase in ^{45}Ca efflux during exposure of vascular smooth muscle to neomycin and interpreted their findings in a similar way. Displacement of calcium ions from binding sites can also account for the reduction in ^{45}Ca uptake observed in our experiments with streptomycin as well as those by Goodman et al. (1974) with neomycin.

RELEASE OF VASOPRESSIN FROM THE NEUROHYPOPHYSIS

At neuromuscular junctions, the main role of calcium is to mediate between presynaptic depolarization and transmitter liberation. In this respect the myoneural junctions resembles other synapses as well as neurosecretory endings (Rahamimoff, 1970).

It has recently been shown that neomycin inhibits both spontaneous and potassium-evoked release of vasopressin from the isolated neurohypophysis of the rat, this effect being accompanied by a decrease in the rate of ^{45}Ca efflux (Batalla-Sotello, 1974). The discrepancy between those findings and the results of our experiments with smooth muscle may reflect differences in the distribution and binding of calcium ions in the two preparations. In the smooth muscle of the guinea pig ileum, the activator calcium ions are associated with a superficially located pool that rapidly equilibrates with the external medium (rate constant: 0.14; Hurwitz and Joiner, 1970). Displacement of calcium ions from that pool would account for the effects of streptomycin as observed in the study of competitive antagonism between calcium and aminoglycoside on smooth muscle. In contrast, the inhibitory effect of neomycin on ^{45}Ca efflux in the rat pituitary might reflect the absence of a similar calcium pool or, alternatively, that calcium ions are more tightly bound. Reduction in membrane calcium permeability could then explain the parallel decrease in vasopressin secretion and ^{45}Ca efflux (Batalla-Sotello, 1974).

There is evidence that neomycin and streptomycin also inhibit the activation of membrane calcium conductance during membrane depolarization in crab skeletal muscle fiber, this effect being attributed to displacement of the relationship between calcium conductance and membrane potential toward higher values of membrane depolarization (Suarez-Kurtz, 1974). A similar mechanism could be involved in the inhibitory action of aminoglycoside antibiotics on stimulus-secretion coupling as well as in the blockade of the calcium evoked contractions in the depolarized guinea pig ileum.

CALCIUM-AMINOGLYCOSIDE ANTAGONISM
IN MEMBRANE MODELS

For this study (Piccinini, Favalli, Chiari, and Corrado, 1973) the three-phase partition model system was used (Rosano, Schulman, and Weisbuch, 1960), because it behaves like a physiological membrane with respect to the different rates of transfer of various inorganic cations (Villani, Chiarra, Peluchetti, and Piccinini, 1969) and because the two interfaces are equivalent to monomolecular phospholipidic layers. The model system allowed us to study the facilitated diffusion of ions from one water compartment to the other (Piccinini, Chiarra, and Villani, 1972) across a phospholipidic layer consisting of a 3.5-mg/ml solution of brain cephalin in a diffusion-blocking solvent, namely, 1-pentanol-petrol ether, 4.1. Salt solutions (0.17 M) of KCl and $MgCl_2$ were allowed to exchange against 0.17 M NaCl and $CaCl_2$, which were placed in the opposite water compartment. The effect of aminoglycoside was studied by using different concentrations in the aqueous compartment containing the calcium ion. The ion concentrations on opposite sides were measured at different time intervals by an Atomic Absorption Spectrometer. The pH value was set at 5 with 0.01 M piperazine buffer.

In the absence of cephalin, a negligible passage of ions occurs by simple diffusion. The addition of cephalin to the diffusion-blocking solvent system brings about a strong increase in the rate of transfer of the four cations studied. Thus we can assume that cephalin is able to act as a nonspecific carrier for both monovalent and divalent cations and that the experimental membrane works as a facilitated diffusion model system.

All antibiotics tested inhibit the transfer of the four ions with the following relative potencies: gentamycin > aminosidin = neomycin > streptomycin. This inhibitory effect is dose dependent, i.e., proportional to the concentration of antibiotic used. Phospholipid ability to carry cations could be blocked by aminoglycoside binding itself in the cationic form to the acidic group of the phospholipid, probably with a higher binding affinity than inorganic monovalent and divalent cations.

Among the phospholipids present in the total extract of cephalin, phosphatidylinositol is selectively active in binding calcium ions with a decreased affinity as follows: Ca > Mg > K > Na (Piccinini, Villani, and Chiarra, 1969). It is interesting to note that aminoglycosides contain a modified inositol moiety in their structure, and it is tempting to speculate that this structural relationship is the basis for their toxic action. It has been suggested that this fraction, acting as a cation, could interact with the calcium-binding sites or, alternatively, could compete with phosphatidylinositides in relation to calcium binding (Corrado, Pimenta de Morais, Prado, and Marseillan, 1971). In agreement with the latter suggestion is the recent report that neomycin inhibits calcium binding and turnover of membrane phosphoinositides from the inner ear tissue of guinea pigs chronically treated with that antibiotic (Schacht, 1974). The possibility that chronic toxicity induced by neomycin can be brought about by inhibition of phosphoinositide turnover and calcium binding is most interesting: it points to a common denominator for the mechanisms of acute and chronic toxicity induced by aminoglycosides and it opens up a new perspective in the prevention and treatment of chronic toxicity of such antibiotics.

ACKNOWLEDGMENTS

We are grateful to Prof. A. Antonio for his many helpful suggestions and criticisms.

This work was supported in part by research grants from the São Paulo State Research Foundation (FAPESP). W. A. P. holds a fellowship from FAPESP. I. P. de M. holds a fellowship from the Foundation of the Brazilian Association of Pharmaceutical Industries (Fundação ABIF), Rio de Janeiro.

REFERENCES

Albert, A. (1965) *Selective toxicity*. Methuen & Co. Ltd., London.

Arunlakshana, O., and Schild, H. O. (1959): Some quantitative uses of drug antagonists. *Br. J. Pharmacol.* 14:48–58.

Batalla–Sotello, L. (1974): Efectos de la neomicina sobre movimientos de calcio y liberación de vasopressina en neurohipofisis de rata. *Thesis,* Faculty of Medicine, University of Rio de Janeiro, Brazil.

Benz, H. G., Lunn, I. N., and Foldes, F. F. (1961): Recurarization by intraperitoneal antibiotics. *Br. Med. J.* 2:241–242.

Bülbring, E. (1946): Observations on the isolated phrenic nerve diaphragm preparation of the rat." *Br. J. Pharmacol. Chemother.* 1:38–61.

Bush, G. H. (1961): Prolonged neuromuscular block due to intraperitoneal streptomycin. *Br. Med. J.* 1:557–558.

Case Report No. 190 (1957): *Am. Soc. Anesthesiologists Newsletter* 21:1:38–40.

Corrado, A. P. (1963): Respiratory depression due to antibiotics: calcium in treatment. *Anesth. Analg. Curr. Res.* 42:1–5.

Corrado, A. P., Pimenta de Morais, I., Prado, W. A., and Marseillan, R. F. (1971): Antagonismo competitivo y no competitivo calcio-antibióticos. In: *Recientes Adelantos en Biologia,* edited by R. H. Mejia and J. A. Moguilevsky, pp. 385–391, Buenos Aires.

Corrado, A. P., and Ramos, A. O. (1958): Neomycin—its curariform and ganglioplegic action. *Rev. Bras. Biol.* 18:81–85.

Corrado, A. P., and Ramos, A. O. (1960): Some pharmacological aspects of a new antibiotic —kanamycin. *Rev. Bras. Biol.* 20:43–50.

Corrado, A. P., Ramos, A. O., and Escobar, C. T. (1959): Neuromuscular blockade by neomycin. Potentiation by ether anesthesia and D-tubocurarine and antagonism by calcium and Prostigmine. *Arch. Int. Pharmacodyn.* 121:380–394.

Del Castillo, J., and Engbaek, L. (1954): The nature of the neuromuscular block produced by magnesium. *J. Physiol.* 124:370–384.

Del Castillo, J., and Katz, B. (1954a): The effect of magnesium on the activity of motor nerve endings. *J. Physiol.* 124:553–559.

Del Castillo, J., and Katz, B. (1954b): Quantal components of the end-plate potential. *J. Physiol.* 124:560–573.

Del Castillo, J., and Stark, L. (1952): The effect of calcium ions on the motor end-plate potentials. *J. Physiol.* 116:507–515.

Dodge, F. A., Jr., and Rahamimoff, R. (1967): Cooperative action of calcium ions in transmitter release at the neuromuscular junction. *J. Physiol.* 193:419–432.

Elmqvist, D., and Josefsson, J.-O. (1962): The nature of neuromuscular block produced by neomycin. *Acta Physiol. Scand.* 54:105–110.

Engel, M. L., and Denson, J. S. (1957): Respiratory depression due to neomycin. *Surgery* 42:862–864.

Fatt, P., and Katz, B. (1951): An analysis of the end-plate potential recorded with an intracellular electrode. *J. Physiol.* 115:320–370.

Fatt, P., and Katz, B. (1952): Some problems of neuromuscular transmission. *Cold Spring Harbor Symposia on Quantitative Biology* 17:275–280.

Ferrara, B. E., and Phillips, R. D. (1957): Respiratory arrest following administration of intraperitoneal neomycin. *Am. Surgeon* 23:710–712.

Fisk, G. C. (1961): Respiratory paralysis after a large dose of streptomycin. *Br. Med. J.* 1:556–557.

Foldes, F. F. (1963): Prolonged respiratory depression caused by drug combinations. Muscle relaxants and intraperitoneal antibiotics as etiological agents. *J.A.M.A.* 183:672–673.

Goodman, F. R., Weiss, G. B., and Adams, H. R. (1974): Alterations by neomycin of ^{45}Ca movements and contractile responses in vascular smooth muscle. *J. Pharmacol. Exp. Ther.* 188:472–480.

Hurwitz, L., and Joiner, P. D. (1969): Excitation-contraction coupling in smooth muscle. *Fed. Proc.* 28:1629–1633.

Hurwitz, L., and Suria, A. (1971): The link between agonist action and response in smooth muscle. *Annu. Rev. Pharmacol.* 11:303–326.

Jenkinson, D. H. (1957): The nature of the antagonism between calcium and magnesium at the neuromuscular junction. *J. Physiol.* 138:434–444.

Jones, W. P. G. (1959): Calcium treatment for ineffective respiration resulting from administration of neomycin. *J.A.M.A.* 170:943–944.

Kownacki, V. P., and Serlin, D. (1960): Intraperitoneal neomycin as a cause of apnea." *A.M.A. Arch. Surg.* 81:838–841.

McCorkle, R. G. (1958): Neomycin toxicity: A case report. *Arch. Pediat.* 75:439–440.

Middleton, W. H., Morgan, D. D., and Moyers, J. (1957): Neostigmine therapy for apnea occurring after administration of neomycin. *J.A.M.A.* 165:2186–2187.

Molitor, H., and Graessle, O. E. (1950): Pharmacology and toxicology of antibiotics. *Pharmacol. Rev.* 2:1–60.

Mullet, R. D., and Keats, A. S. (1961): Apnea and respiratory insufficiency after intraperitoneal administration of kanamycin. *Surgery* 49:530–533.

Pandey, K. (1964): Neuromuscular blocking and hypotensive actions of streptomycin and their reversal. *Br. J. Anaesth.* 36:19–25.

Piccinini, F., Chiarra, A., and Villani, F. (1972): The active form of local anesthetic drugs. *Experientia* 28:140–141.

Piccinini, F., Favalli, L., Chiari, M. C., and Corrado, A. P. (1973): Interferência induzida por antibióticos aminoglicosídeos—aminociclitólicos no transporte Iônico em modelos de membrana. *Ciênc. Cult.* 25:493.

Piccinini, F., Villani, F., and Chiarra, A. (1969): Comparazione di attività di alcuni "carriers" fosfolipidici sul transporto verso un solvente organico di cationi mono e bi-valenti. Effetto di alcuni farmaci. *Atti Accad. Med. Lomb.* 24:80–84.

Pittinger, C. B., and Adamson, R. (1972): Antibiotic blockade of neuromuscular function. *Am. Rev. Pharmacol.* 12:169–184.

Pittinger, C. B., and Long, J. O. (1958): The neuromuscular blocking action of neomycin sulfate. *Antib. Chemother.* 8:198–203.

Pridgen, J. E. (1956): Respiratory arrest thought to be due to intraperitoneal neomycin. *Surgery* 40:571–574.

Rahamimoff, R. (1970): Role for calcium ions in neuromuscular transmission. In *Calcium and Cellular Function,* edited by A. W. Cuthbert, pp. 131–147. MacMillan and Co. Ltd., London.

Richet, G., Ducrot, H., and Delzant, J. F. (1960): Les grands accidents des antibiotiques. *La Presse Médicale* 68:11–12.

Rocha e Silva, M. (1959): Concerning the theory of receptors in pharmacology. A rational method of estimation of pA_x. *Arch. Int. Pharmacodyn.* 118:74–94.

Rosano, H. L., Schulman, J. H., and Weisbuch, J. B. (1960): Mechanism of the selective flux of salts and ions through monaqueous liquid membranes. *Ann. N.Y. Acad. Sci.* 92:457–469.

Schacht, J. (1974): Interaction of neomycin with phosphoinositide metabolism in guinea pig inner ear and brain tissues. *Ann. Otol. Rhinol. Laryngol.,* 83:613–618.

Schild, H. O. (1947): pA: A new scale for the measurement of drug antagonism. *Br. J. Pharmacol.* 2:189–206.

Suarez-Kurtz, G. (1974): Inhibition of membrane calcium activation by neomycin and streptomycin in crab muscle fibers. *Pfluegers Arch.* 349:337–349.

Villani, F., Chiarra, A., Peluchetti, D., and Piccinini, F. (1969): Determinazioni di transporto ionico in un modello di membrana in presenza o no di "Carriers." Influsso di un anestetico locale. *Atti Accad. Med. Lomb.* 24:71–80.

Vital Brazil, O. (1961): Streptomycin effect on the skeletal muscle stimulation produced by acetylcholine. *Arch. Int. Pharmacodyn.* 130:136–140.

Vital Brazil, O., and Corrado, A. P. (1957): The curariform action of streptomycin. *J. Pharmacol. Exp. Ther.* 120:452–459.

Vital Brazil, O., Corrado, A. P., and Berti, F. A. (1957): Neuromuscular blockade produced by streptomycin and some of its degradation products. In: *Curare and Curare-like Agents,* edited by D. Bovet, F. Bovet–Nitti, and G. B. Marini–Bettolo, pp. 415–421. Elsevier, Amsterdam.

Webber, B. M. (1957): Respiratory arrest following intraperitoneal administration of neomycin. *Arch. Surg.* 75:174–176.

Concepts of Membranes in Regulation and Excitation,
edited by M. Rocha e Silva and G. Suarez-Kurtz.
Raven Press, New York © 1975.

Biochemical Properties of Tityus Toxin

C. R. Diniz, J. Coutinho Netto, A. F. Pimenta, and R. E. Larson

Laboratory of Neurochemistry, Department of Biochemistry, Faculty of Medicine of Ribeirão Preto, University of São Paulo, 14100 Ribeirão Preto, S.P. Brazil

The venoms from the South American scorpions *Tityus serrulatus* (Lutz and Mello, 1922) and *Tityus bahiensis* (Perty, 1860) were shown by Diniz and Gonçalves (1960), employing electrophoretic techniques, to be composed primarily of basic proteins. The toxic components so separated were found to contract the smooth muscle of guinea pig ileum. Unlike snake venoms, the scorpion venoms are relatively devoid of enzymatic activity. Consequently, when a mammalian host is stung, there is little or no necrosis at the site of sting, and toxicity is credited to an action upon the nervous structures.

PURIFICATION OF TITYUS TOXIN

Initial attempts to prepare toxic components from *Tityus* venom by chromatographic techniques were made by Gomez and Diniz (1966) and Miranda, Rochat, Rochat, and Lissitzki (1966). These authors separated two active components in the venom by column chromatography using Sephadex G-25. One of the components (T_1) was further purified by Gomez and Diniz (1966), and Gomez (1971). A highly active toxin—tityus toxin (TsTx) (LD_{50} 125 ug/kg mouse)—was obtained, which was found to be homogeneous by acrylamide gel electrophoresis and chromatography. The determination of the amino terminal residue indicated only one amino acid—lysine.

More recently a new method of purification of TsTx has been worked out in our laboratory, employing gel permeation with Sephadex G-50 and ammonia elution. The results indicate that the toxic component in the venom is among the proteins of lowest molecular weight. Further purification of the toxic component by equilibrium chromatography in CM-Cellulose–W-52 columns at pH 8.5 and 0.15 M ammonium acetate leads to the isolation of a basic protein (p.I.8.25) that is homogeneous to polyacrylamide gel electrophoresis. Amino acid analysis showed the following composition: Lys_7, His_1, Arg_1, Cys_6 (as $CySO_3$), Asp_9, $Meth_0$, Thn_2, Ser_3, Glu_4, Pro_3, Gly_5, Ala_4, Val_2, Ile_2, Leu_3, Tyr_6, Phe_1, Trp_2. Minimum molecular weight is 6,995 daltons, Lysine is the N terminal amino acid. The number of hydrophobic residues showed 16 out of a total of 61 residues.

TsTx-STIMULATED RELEASE OF NEUROTRANSMITTERS

The observations that *Tityus* venoms cause contraction of muscular organs such as guinea pig ileum and heart and that those effects are influenced by drugs that antagonize or potentiate neurotransmitters led to the hypothesis that the contractile effects were essentially due to the release of chemical mediators from the nerve terminals in these organs (see Corrado, Diniz, and Antonio, *this volume*). The discovery that both the venom and TsTx can stimulate the release of acetylcholine from the guinea pig ileum and brain slices (Gomez, Dai, and Diniz, 1973) and the release of norepinephrine from the heart atria (Langer, Adler Grachinsky, Alméida, and Diniz, *unpublished results*) supports the hypothesis.

Venom of other species of scorpions probably acts in a similar way. Re-

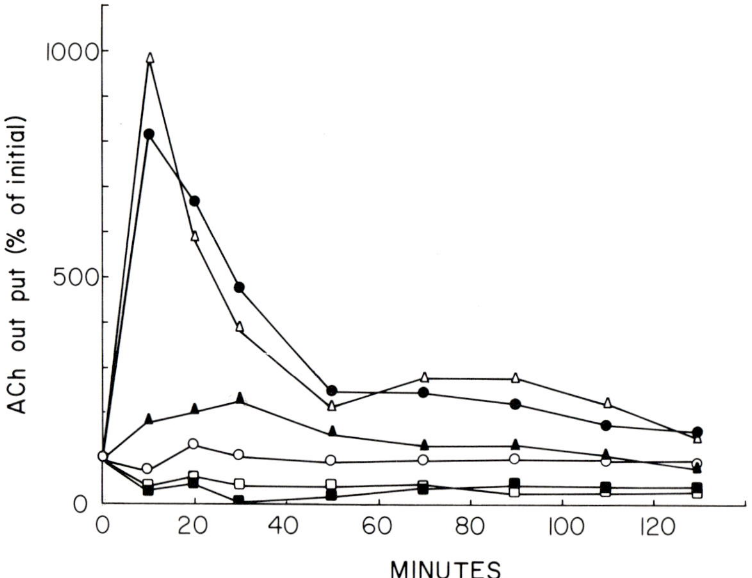

FIG. 1. Effect of TsTx on acetylcholine (ACh) output from eserinized longitudinal muscle strip of guinea pig ileum. Initial rate of ACh output in presence of normal Krebs solution, 0.23 ± 0.03 nmoles/g/min taken as 100%. TsTx added as indicated in the text to reach a concentration of 12 μg/ml. ●—●, ACh release of a strip exposed continuously to TsTx in normal Krebs solution. △—△, Strip incubated only in the first 10 min with TsTx; during subsequent periods strip was suspended in normal Krebs solution. ▲—▲, Rate of ACh release in a strip treated with TsTx in a Na-free solution, sucrose substitution. □—□, Output of ACh produced by TsTx in strips bathed in a Ca-free Krebs solution. ■—■, TTx and TsTx added simultaneously to the normal bathing Krebs solution. ○—○, Output of ACh of control strips suspended in normal Krebs solution in absence of TsTx. Output of control strips in Na-free and Ca-free solutions or exposed to TTx in absence of TsTx were not shown and did not differ significantly from the TsTx-exposed strips in these conditions. Each point represents the ACh output during the preceding period and was the mean of two experiments on the same schedule (A. F. Pimenta and C. R. Diniz, *unpublished results*).

cently Moss, Thoa, and Kopin (1974) described the effects of a purified fraction of the venom from *Leiurus quinque striatus,* an African scorpion, on several preparations—namely, brain synaptosomes, heart atria, and vas deferens. The venom of that species releases norepinephrine from the prejunctional structures of the nervous system.

The site of action of the toxin seems most likely to be on the presynaptic membrane, since release of neurotransmitters by the toxin is greatly affected by the ionic environment of the preparations. Absence of Na^+ or Ca^{++} in the incubating fluids prevents the release of acetylcholine by the toxin. Also tetrodotoxin (TTx), a substance known to specifically depress the membrane's inward permeability to sodium ions, blocks the effect of TsTx (Fig. 1).

A new method is being developed in our laboratory to monitor the TsTx-stimulated release of acetylcholine by the longitudinal muscle of guinea pig ileum during the first few minutes in the presence of the toxin. Within 30 sec after the introduction of TsTx, a significant increase in the rate of release, compared to the resting level, can be seen (Fig. 2). Under optimal

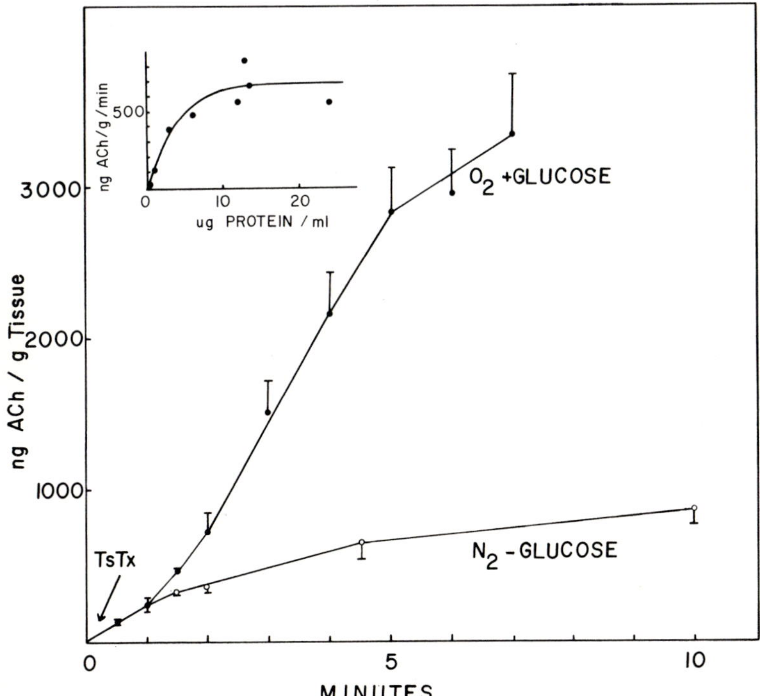

FIG. 2. The release of ACh into the incubation bath by the longitudinal muscle from guinea pig ileum is plotted against the time of incubation with 13 μg/ml of TsTx, in the presence and absence of glucose and O_2. The upper-left-hand insert plots the steady rate of ACh release observed during the first few minutes (usually 2–5 min) in the presence of an increasing concentration of TsTx.

conditions of 12 mM glucose and with bubbling of 95% O_2 plus 5% CO_2 gas, the rate of release reaches a maximum within 1 to 2 min and remains steady for at least 5 to 10 min. In the absence of glucose and with replacement of O_2 by N_2, the TsTx-stimulated release of acetylcholine is dramatically inhibited. Note, however, that this inhibition does not appear until after the first minute of introduction of the toxin. The rate of release is not different up to about the first minute in the presence or absence of glucose and O_2, but is clearly greater than the resting level. O_2 alone is able to alleviate this inhibition to a certain degree (see Table 1) but the rate of release does not reach the maximum observed in the presence of glucose and O_2. Preincubation of the longitudinal muscle in the presence of metabolic inhibitors (iodoacetic acid, sodium azide, dinitrophenol, and dinitrofluorobenzene) inhibit the TsTx-stimulated release, including the initial 30 sec to 1 min. These results suggest that the release of acetylcholine stimulated by TsTx is dependent on processes requiring metabolic energy.

TABLE 1. *Liberation of ACh (ng/g tissue/min) from longitudinal muscle strip of guinea pig ileum in the presence and absence of glucose and O_2.*

Solution	Incubation without TsTx	Incubation with 13 μg TsTx/ml			
		4 min	7 min	8 min	2–10 min
12 mM glucose + O_2	98 ± 16 (9)	556 ± 67 (7)	480 ± 57 (5)	—	—
Absence of glucose ± O_2	28 ± 3 (8)	271 ± 34 (6)	—	331 (2)	301 ± 25 (11)
Absence of glucose + N_2	52 ± 6 (8)	111 ± 18 (4)	—	124 ± 34 (4)	118 ± 25 (8)

Our studies are not only clarifying the mechanism of action of TsTx in the release of neurotransmitters but are indicating the potential value of this toxin as a tool for studying synaptic events.

ACKNOWLEDGMENTS

This work was supported by research grant 72/1341 from the São Paulo State Research Foundation (FAPESP).

REFERENCES

Diniz, C. R., and Gonçalves, J. M. (1960): Separation of biologically active components from scorpion venoms by one electrophoresis. *Biochim. Biophys. Acta* 41:470–477.

Gomez, M. V. (1971): Dissertation for doctorate degree. Universidade Federal de Minas Gerais, Belo Horizonte.

Gomez, M. V., Dai, M. E., and Diniz, C. R. (1973): Effect of scorpion venom, Tityustoxin, on the release of acetylcholine from incubated slices of rat brain. *J. Neurochem.* 20:1051–1061.

Gomez, M. V., and Diniz, C. R. (1968): Separation of toxic components from the Brazilian scorpion—*Tityus serrulatus*-venom. *Mem. Inst. Butantan* 33:899–902.

Langer, S., Adler-Graschinsky, E., Almeida, A. P., and Diniz, C. R. (1975): Prejunctional effects of a purified toxin from the scorpion Tityus serrulatus. Release of ^{3}H-noradrenaline and enhancement of transmitter overflow elicited by nerve stimulation. *Naunyn Schmiedebergs Arch. Pharmacol. In press.*

Miranda, F., Rochat, H., Rochat, C., and Lissitzky, S. (1966): Chromatographic purification of the venom neurotoxins of a South American scorpion (*Tityus serrulatus*). *Toxicon* 4:145–152.

Moss, J., Thoa, N. B., and Kopin, I. J. (1974): On the mechanism of scorpion toxin-induced release of norepinephrine from peripheral adrenergic neurons. *J. Pharmacol. Exp. Ther.* 190:39–48.

SUBJECT INDEX

Acetazolamide
    carbonic anhydrase and, 107-108
    mechanism of action, 117-122
    renal acidification and, 117-122
    transport and, 116
    urinary bladder and, 128-130
Acetylcholine (ACh)
    receptors for, 60-64
    release of, 193-195, 203, 218-220
    scorpion venom and, 193-195
    smooth muscle contraction and,
        60-64
    synthesis, 198
    Tityus toxin and, 218-220
    vas deferens contractions and, 73
Adenosine triphosphate (ATP)
    calcium regulation of, 16-17
    calcium uptake and, 7-17
    exchange of, 9-17
    hydrolysis of, 11-14
    inorganic phosphorus and, 9-17
    membrane phosphorylation and,
        16-17
    sarcoplasmic reticulum and, 7-17
    silver ions and, 14-15
    sodium transport and, 104-105
Adenyl cyclase, in skeletal muscle, 3
Adrenergic receptors, in smooth
      muscle, 62
Aminoglycoside antibiotics
    Ca action and, 51-52, 201-213
    in membrane models, 212-213
    neuromuscular junction and, 203
    structures of, 201-203
    vasopressin and, 211
Anesthetics, malignant hyperthermia
      and, 4-5
Angiotensin
    binding studies of, 173
    calcium liberation and, 145-146
    characterization of, 146-147
    desensitization to, 147-148
    function of, 145-146

    plasma membrane and, 174-177
    receptors
        characterization of, 146-147
        isolation of, 146-147
        labeling of, 152-153
    smooth muscle and, 145-153
    spin probes of, 174
    structure of, 146-147
ATPase
    Ca activation of, 170-172
    calcium gradient and, 9
    leaky vesicles and, 8-9, 16-17
    of smooth muscle, 170-172

Barium, contractions of, 73, 79-82
Benadryl, drug antagonism and, 181,
      185
Bicarbonate reabsorption, 109-111

Caffeine
    calcium fluxes and, 1, 3-5, 36-37
    cyclic AMP and, 3
    Mg and, 31-32
    skeletal muscle and, 3-5, 29-30,
        33-34
    "skinned" muscle and, 29-30,
        33-34, 36
Calcium
    acetylcholine receptor and, 148
    angiotensin and, 145-153
    antibiotics and, 51-52, 201-213
    ATP exchange and, 9-17
    ATPase and, 170-172
    Ca release and, 34-36
    caffeine and, 33-34, 36
    conductance, 46-47
    crab fibers and, 41-52
    drug receptor complex and, 57-64,
        73-74
    excitation process and, 41-52,
        55-57
    leaky vesicles and, 8-9, 16-17
    membrane depolarization and, 1-2

membrane permeability and, 135-
136
mobilization of, 65-69
neuromuscular junction and, 203-
213
oxygen uptake and, 1-5
pumps for, 68
regulation of, 21-37, 64
reservoirs of, 65-69
sarcoplasmic reticulum and, 7-17
sequestration of, 23-29, 75
silver ions and, 14-15
in skeletal muscle, 1-5
in smooth muscle, 55-70
spike thresholds and, 42-45
tension development and, 27-30,
42-43
transport of, 55-57, 69-70
troponin and, 36-37
uncoupling and, 141-142
vasopressin and, 211
Versene and, 209-210
Cancer cells, growth control in, 136
Carbonic anhydrase
function of, 107-109
inhibition of, 107-109, 116, 121
in renal tubules, 107-109
sodium chloride transport and, 116
in urinary bladder, 125
Cardiac muscle
action potentials of, 90-92
phase-plane analysis of, 88-90
scorpion venom and, 195-198
sodium ions and, 90-92
Communication, intercellular, 135-
142
Crab muscle, excitation in, 41-52
Cyanide, sodium transport and,
101-102
Cyclic AMP, in skeletal muscle, 3
Cyclic GMP, in skeletal muscle, 3

Dibenamine, muscle receptors and,
58, 62-63
Dibenzyline, antagonism and, 185-
186
2,4-Dinitrophenol, sodium transport and, 101
Drug antagonism
to Benadryl
Charnière effect, 181-186
to histamine, 181
kinetics of, 182-191

Drug antagonism, *contd.*
local anesthetics, 182
models of, 186-191
Drug-receptor complex
acetylcholine and, 60-64
adrenergic receptors and, 62
Ca transport and, 57-64, 73-74
excitation of, 57-64
in smooth muscle, 57-64
spare receptors and, 62
in vas deferens, 73-82

Electron microscopy
of sarcolemma, 168-169
of smooth muscle, 168-169
Epinephrine
adenyl cyclase and, 3
calcium fluxes and, 2-3
smooth muscle and, 62
vas deferens and, 73, 79
Ethacrynic acid, carbonic anhydrase and, 128-131
Excitation-contraction coupling
Ca current and, 41-52
drug-receptor interaction in,
57-64
hypothesis for, 55-57

Furosemide, carbonic anhydrase and,
128-132

Histamine
antagonism, 181
receptors of, 63-64
Hydrogen ion transport
acetazolamide effect, 107-108,
117-122
bicarbonate reabsorption and,
109-114
carbonic anhydrase and, 107-109
kinetics of, 114-116
pH recording in, 111-114
in renal tubules, 107-122
in urinary bladder, 125-132

Kenamycin
neuromuscular junction and, 204-205
smooth muscle and, 207-208
structure of, 203

Lipid bilayers
permeability of, 155-156, 161-
163

Lipid bilayers, *contd.*
    preparation of, 157-158
    spin probes of, 156-163
Lymphocytes
    antibody response in, 136
    cell junctions in, 137-139
    communications in, 135
    culture of, 137
    phytohemagglutinin and, 135-139
    stimulation mechanisms, 136-137
    transjunctional permeability in,
        136-142

M agnesium
    muscle tension and, 31-33
    neuromuscular junction and, 204-
        206
    troponin and, 36
Malignant hyperthermia, 4-5
Manganese, Ca conductance and, 47,
    52
Membrane currents, computation of,
    86-88
Membrane potential
    calcium fluxes and, 1-3
    in skeletal muscle, 1-2
Membranes
    antibodies and, 212-213
    communications across, 135-136
    composition of, 169-170
    current recordings, 86-90
    hydrogen ion transport in, 107-122
    lipid bilayer, 155-163
    oxygen coupling in, 97-104
    permeability of, 135-136
    plasma, 167-173, 175-177
    sodium transport in, 97-104
Muscle tension
    caffeine and, 29-30
    calcium and, 27-30, 42-49

N eomycin
    Ca activation and, 50-52
    Ca binding and, 211
    structure of, 201-203
Norepinephrine
    release by venom, 197-198
    scorpion venom and, 195-198

O uabain, sodium transport and, 100
Oxygen uptake
    adenyl cyclase and, 3,5
    in amphibian skin, 99-104

Oxygen uptake, *contd.*
    calcium and, 1-5
    drug effects, 100-102
    in skeletal muscle, 1-5
    sodium transport and, 99-100

P hase-plane recording
    of action potentials, 85-86
    of atrium, 88-90
    of cardiac muscle, 93-94
    of membrane characteristics,
        88-90
Phytohemagglutinin, lymphocyte
        stimulation of, 135-139
Procaine
    calcium fluxes and, 2
    tension development and, 31-33

R eceptor, *see* Drug-receptor
        complex
Receptors
    for acetylcholine, 148, 193-194
    for antiotensin, 145-153
Renal tubules
    carbonic anhydrase and, 107-109
    hydrogen ion sources in, 119-122
    hydrogen ion transport in, 107-122
    microperfusion studies, 111-114

S arcoplasmic reticulum
    ATP and, 7-17
    calcium uptake and, 9-17
    silver ions and, 14-15
    in "skinned" muscle, 36-37
    vesicles, 8-17
Scorpion venom
    acetylcholine release by, 193-194,
        109
    adrenergic actions of, 195-198
    cholinergic actions, 193-195
    hypertensive response to, 197-198
    Tityus toxin, 217-220
Silver ions, calcium uptake and, 14-15
Skeletal muscle
    adenyl cyclase in, 3
    aminoglycoside antibiotics and,
        201-213
    ATP exchange in, 7-17
    Ca mobilization in, 65
    calcium fluxes in, 1-5
    cyclic AMP and, 3
    glycerol-treated, 2-3
    membrane potential in, 1-2

Skeletal muscle, *contd.*
  phase-plane recordings in, 93-94
  skinned fibers of, 21-37
"Skinned" muscle fibers
  anion effects, 30-31
  Ca regulation in, 21-37
  Ca release and, 34-36
  Ca uptake in, 23-27
  caffeine and, 29-30, 33-34
  magnesium and, 31-33
  preparation of, 21-23
  tension development in, 27-32
Smooth muscle
  acetylcholine and, 60-64
  adrenergic receptors in, 62
  aminoglycoside antibiotics and,
    201-205, 207-211
  angiotensin receptors in, 145-153,
    167-178
  ATPase of, 170-172
  Ca ions and, 55-57
  Ca mobilization in, 65-69
  drug-receptor interaction in, 57-64
  excitation-contraction in, 55-70
  histamine receptors in, 63-64
  membranes
    chemical composition, 169-170
    enzymes of, 170-172
  scorpion venom and, 193-194
  tachyphylaxis of, 147-152
Sodium chloride transport, 116
Sodium ions, in cardiac muscle and,
    90-92
Sodium transport
  in amphibian skin, 97-104
  drug effects, 100-101
  oxygen coupling, 99-104
  potential differences and, 102-104
  in urinary bladder, 128
Spin probes
  activation energies of, 160-162
  of angiotensin receptors, 146, 174
  kinetics of , 158-160
  of lipid bilayers, 156-157
  of model systems, 177-178
  spectra of, 159
Streptomycin
  Ca binding and, 210-211
  structure of, 202
Strontium, smooth muscle and, 65

T achyphylaxis
  angiotensin and, 147-152

Tachyphylaxis, *contd.*
  pH and, 148-150
  receptor and, 148-152
Tityus toxin
  neurotransmitter release by, 218-
    220
  properties of, 217
  purification of, 217
Transjunctional permeability, in
    lymphocytes, 136-142
Troponin
  Ca and, 36-37
  Mg and, 36

U rinary bladder
  acidification in, 125-132
  carbonic anhydrase of, 125, 128-
    130
  natriuretic agents in, 128-130.
  tubular preparations of, 126-127

V as deferens
  ACh contractions in, 73
  barium contractions in , 73, 79-82
  Ca release in, 75-79
  Ca transfer in, 73-74
  epinephrine effects, 73-79
  receptor mechanisms in, 73-82
Vasopressin, Ca binding and, 211